COMPARING VOTING SYSTEMS

THEORY AND DECISION LIBRARY

General Editors: W. Leinfellner and G. Eberlein

Series A: Philosophy and Methodology of the Social Sciences
Editors: W. Leinfellner (Technical University of Vienna)
G. Eberlein (Technical University of Munich)

Series B: Mathematical and Statistical Methods
Editor: H. Skala (University of Paderborn)

Series C: Game Theory Mathematical Programming and Mathematical Economics
Editor: S. Tijs (University of Nijmegen)

Series D: System Theory, Knowledge Engineering and Problem Solving
Editor: W. Janko (University of Vienna)

SERIES A: PHILOSOPHY AND METHODOLOGY OF THE SOCIAL SCIENCES

Editors: W. Leinfellner (Technical University of Vienna)
G. Eberlein (Technical University of Munich)

Editorial Board

M. Bunge (Montreal), J. S. Coleman (Chicago), M. Dogan (Paris), J. Elster (Oslo), L. Kern (Munich), I. Levi (New York), R. Mattessich (Vancouver), A. Rapoport (Toronto), A. Sen (Oxford), R. Tuomela (Helsinki), A. Tversky (Stanford).

Scope

This series deals with the foundations, the general methodology and the criteria, goals and purpose of the social sciences. The emphasis in the new Series A will be on well-argued, thoroughly analytical rather than advanced mathematical treatments. In this context, particular attention will be paid to game and decision theory and general philosophical topics from mathematics, psychology and economics, such as game theory, voting and welfare theory, with applications to political science, sociology, law and ethics.

HANNU NURMI

*Department of Political Science,
University of Turku, Finland*

COMPARING VOTING SYSTEMS

D. REIDEL PUBLISHING COMPANY

A MEMBER OF THE KLUWER ACADEMIC PUBLISHERS GROUP

DORDRECHT / BOSTON / LANCASTER / TOKYO

Library of Congress Cataloging in Publication Data

Nurmi, Hannu.
    Comparing voting systems.

    (Theory and decision library. Series A, Philosophy and methodology of the social sciences)
    Bibliography; p.
    Includes indexes.
    1. Political science—Decision making.   2. Voting.   3. Game theory.
4. Social choice.   I. Title.   II. Series.
JA74.N86 1987   320'.072   87-20807
ISBN 90-277-2600-0

Published by D. Reidel Publishing Company,
P.O. Box 17, 3300 AA Dordrecht, Holland.

Sold and distributed in the U.S.A. and Canada
by Kluwer Academic Publishers,
101 Philip Drive, Assinippi Park, Norwell, MA 02061, U.S.A.

In all other countries, sold and distributed
by Kluwer Academic Publishers Group,
P.O. Box 322, 3300 AH Dordrecht, Holland.

All Rights Reserved
© 1987 by D. Reidel Publishing Company, Dordrecht, Holland
No part of the material protected by this copyright notice may be reproduced or
utilized in any form or by any means, electronic or mechanical
including photocopying, recording or by any information storage and
retrieval system, without written permission from the copyright owner

Printed in The Netherlands

TABLE OF CONTENTS

PREFACE ix

CHAPTER 1. INTRODUCTION 1

CHAPTER 2. PRELIMINARIES 5

CHAPTER 3. SOCIAL WELFARE FUNCTION, SOCIAL CHOICE FUNCTION AND VOTING PROCEDURES 8

CHAPTER 4. FIRST PROBLEM: CYCLIC MAJORITIES 12

- 4.1. The Condorcet paradox 12
- 4.2. How to conceal the problem: the amendment procedure 14
- 4.3. How common are the cycles 16
- 4.4. Solutions based on ordinal preferences 24
    - 4.4.1. Schwartz' procedure 25
    - 4.4.2. Dodgson's procedure 26
    - 4.4.3. The maximin method 29
- 4.5. Solution based on scoring function: the Borda count 31
- 4.6. More general majority cycles 34

CHAPTER 5. SECOND PROBLEM: HOW TO SATISFY THE CONDORCET CRITERIA 38

- 5.1. Condorcet criteria 38
- 5.2. Some complete successes 40
- 5.3. Some partial successes 47
- 5.4. Complete failures 56
- 5.5. Some probability considerations and the plausibility of the Condorcet criteria 58
- 5.6. The majority winning criterion 62

CHAPTER 6. THIRD PROBLEM: HOW THE AVOID PERVERSE RESPONSE TO CHANGES IN INDIVIDUAL OPINIONS 66

- 6.1. Monotonicity and related concepts 66

6.2.	Successes	69
6.3.	Failures	74
6.4.	The relevance of the monotonicity criteria	78

CHAPTER 7. FOURTH PROBLEM: HOW TO HONOUR UNANIMOUS PREFERENCES 81

7.1.	Unanimity and Pareto conditions	81
7.2.	Successes	83
7.3.	A partial failure and a total failure	85
7.4	Relevance and compatibility with other criteria	87

CHAPTER 8. FIFTH PROBLEM: HOW TO MAKE CONSISTENT CHOICES 92

8.1.	Choice set invariance criteria	92
8.2.	Performances with respect to consistency	100
8.3.	Performances with respect to WARP and PI	104
8.4.	The relevance of the criteria	106

CHAPTER 9. SIXTH PROBLEM: HOW TO ENCOURAGE THE SINCERE REVELATION OF PREFERENCES 108

9.1.	Manipulability	109
9.2.	Performance with respect to manipulability	112
9.3.	The difficulty of manipulation	118
9.4.	Agenda-manipulability	125
9.5.	Sincere truncation of preferences	128

CHAPTER 10. SOCIAL CHOICE METHODS BASED ON MORE DETAILED INFORMATION ABOUT INDIVIDUAL PREFERENCES 135

10.1.	The von Neumann-Morgenstern utility and classes of interpersonal comparability	136
10.2.	Old and new methods	140
10.3.	An assessment	143

CHAPTER 11. ASKING FOR LESS THAN INDIVIDUAL PREFERENCE ORDERINGS 148

11.1.	Constructing a social preference order for a subset of alternatives	149
11.2.	Results based on individual choice functions	155

TABLE OF CONTENTS vii

CHAPTER 12. WHY IS THERE SO MUCH STABILITY
AND HOW CAN WE GET MORE OF IT? 160

    12.1. Explanations of stability 160
    12.2. Improving the performance of the voting procedures 172

CHAPTER 13. FROM COMMITTEES TO ELECTIONS 178

    13.1. Proportional and majoritarian systems 178
    13.2. Criteria for proportional systems 181
    13.3. Voting power 184

CHAPTER 14. CONCLUSIONS 191

BIBLIOGRAPHY 194

SUBJECT INDEX 202

NAME INDEX 207

PREFACE

In many contexts of everyday life we find ourselves faced with the problem of reconciling the views of several persons. These problems are usually solved by resorting to some opinion aggregating procedure, like voting. Very often the problem is thought of as being solved after the decision to take a vote has been made and the ballots have been counted. Most official decision making bodies have formally instituted procedures of voting but in informal groups such procedures are typically chosen in casu. Curiously enough people do not seem to pay much attention to which particular procedure is being resorted to as long as some kind of voting takes place. As we shall see shortly the procedure being used often makes a great difference to the voting outcomes. Thus, the question arises as to which voting procedure is best. This book is devoted to a discussion of this problem in the light of various criteria of optimality. We shall deal with a number of procedures that have been proposed for use or are actually in use in voting contexts. The aim of this book is to give an evaluation of the virtues and shortcomings of these procedures. On the basis of this evaluation the reader will hopefully be able to determine which procedure is optimal for the decision setting that he or she has in mind.

Over the years I have been working on this book I have benefited from the advice, criticism and encouragement of a number of persons. An early version of this book was read and commented upon by Professor Michael Laver. Dr. Gerald Doherty has checked the language of this and many earlier versions. His contribution to this project extends far beyond that of linguistic counsel. An anonymous referee of D. Reidel Publishing Company has made many perceptive comments on the penultimate draft of this book.

Several details of the manuscript have been commented upon by Peter Wakker, Dipankar Ray and John Coakley. I am also grateful to Professors Steven J. Brams, Richard G. Niemi and William H. Riker for discussions and correspondence related to voting systems. My further intellectual debts are easily recognizable from a glance at the references. The support and encouragement of Professors Jaakko Hintikka and Risto Hilpinen during the final stages of the preparation of the manuscript is highly appreciated.

The numerous drafts of this book were typed by Mrs. Katriina Nurmi. The final text-processing was performed by Mrs. Anneli Hilpinen. Financial support was provided by the Academy of Finland and the Foundation of the University of Turku. The members of my family have, above all, helped me to maintain a reasonable sense of proportion concerning the importance of voting systems in the fortunes of mankind. I hereby express my sincere gratitude to all these persons and institutions.

Turku, February 1987   HANNU NURMI

CHAPTER 1

INTRODUCTION

Most of us have participated in elections. Many political offices are determined by voting, but so are the most popular songs in song contests, the sportsmen of the year, the winners of beauty contests etc.. The range of situations in which voting is used to determine which one among a group of individuals is considered as the best candidate is vast. But the situations involving voting have an even wider scope: also policy or action alternatives can be determined by voting. This, indeed, is what the bulk of parliamentary decision making is concerned with.

If the variety of situations involving voting is large, so is the number of different voting procedures used in various voting bodies. In view of the fact that the motivation for the use of the procedures often amounts to the same, viz. that the alternative which the group as a whole regards as best should chosen, one may well wonder why so many procedures are needed for the same purpose. This book centers around this problematique. More specifically, I shall focus on the similarities and differences of various voting procedures with the aim of giving a balanced view of their virtues and vices.

This book in intended mainly for people who design, use or are otherwise concerned with voting procedures as ways of making collective decisions. My intention is to show that different voting procedures are, indeed, different in a very basic sense, viz. the same voting body may come up with different collective decisions when different voting procedures are utilized. And yet, the procedures purport to do the same, i.e. elicit the collective opinion of the best candidate or alternative.

Just how different the procedures can be is illustrated by the following fictitious example. Consider a 9-member voting body assigned to the task of choosing an office-holder from among 5 applicants. The members of the voting body - hereinafter the voters - form three opinion groups consisting of 4, 3 and 2 voters. Each group has an identical opinion about the preference order of the applicants, that is, each group has the same idea of the relative competence of the applicants for the task at hand. Thus, group 1 thinks that of the five applicants a, b, c, d and e, the applicant a is best, e next best, then d, c and finally b. Group 1 consists of 4 members. Group 2, on the other hand, thinks that b is best, followed by c which again is followed by e, then d and finally a. Group 2 consists of 3 voters. Group 3, in turn, has the preference order from the best to the worst: c, d, e, b, a. This group has 2 members.

It is convenient to present the views of the groups as follows:

group 1	group 2	group 3
4 persons	3 persons	2 persons
a	b	c
e	c	d
d	e	e
c	d	b
b	a	a

That is, the most-preferred alternative is the upper-most one, the next best on the next row and so on. This notational convention will be followed throughout the book.

We see that groups 1 and 2 have opposing interests with respect to the ranking of a and b: group 1 considers a as best and b as worst, while group 2 considers b as best and a as worst. Group 3, in turn, likes neither a nor b as they are ranked worst and next to worst by it, respectively.

Now under certain additional and not too implausible assumptions I can show that five voting procedures, each of which is used in some real world voting body, all result in a different choice for the office to be filled in the present example. Consider first the simple plurality procedure which declares the winner to be the alternative which has been given the largest number of votes supposing that such an alternative exists. If we assume that each voter votes for the alternatives she or he (hereinafter he) ranks highest, a will be given 4 votes, b 3 votes and c 2 votes. Hence a wins.

Suppose now that the plurality runoff procedure were used. Under this procedure, an alternative with more than 50 % of the votes will be declared the winner if such an alternative exists. Otherwise, a second ballot is taken between the two largest vote-getters in the first ballot. The alternative with the larger number of votes is then the winner. Assuming again that the voters vote for their highest ranked alternatives, there will be no winner in the first round as none of the alternatives gets 5 votes or more.

INTRODUCTION

The second ballot will then take place between a and b. Obviously groups 1 and 2 will in this round vote for a and b, respectively, while group 3 will presumably vote for b, it being ranked higher than a by this group. Now b will be the winner.

Some voting bodies resort to pairwise comparisons of alternatives in determining the winner. In particular, some procedures will declare the winner the alternative that beats all the others in pairwise comparisons by a simple majority of votes. Of course, such an alternative may not always exist and hence some method of determining the winner must be devised for such a contingency. Under fairly plausible assumptions there will, however, be an alternative that would beat the rest in pairwise comparisons in this example. A glance at the ranking of alternatives in the three groups shows that c would be the victorious one should the pairwise comparison be resorted to. It is ranked first by group 3 and hence will get at least 2 votes in each pairwise comparison against the other alternatives. This, of course, will not suffice, but against alternatives a, d and e group 3 will be joined by group 2 which also ranks c higher than these three alternatives. Together groups 2 and 3 clearly form a majority and c will thus win these three pairwise comparisons. The only remaining pairwise comparison for c is against b. In this comparison group 1 will presumably join hands with group 3 as b is ranked worst by group 1. Thus c beats all pairwise comparisons by a simple majority. As there cannot be more than one alternative which is so victorious in pairwise comparisons, c will be the winner.

Some voting bodies use point counting procedures, like the well-known Borda count. Assume that the number of alternatives is k. In this procedure each voter's highest ranked alternative is given k-1 points, the next-highest k-2 points and so on until the lowest ranked one, which is given 0 points. The points are then summed up for each alternative or candidate and the one with the largest sum score is chosen. In our example group 1 would thus give a 4 times 4 or 16 points, while the others would give it 0 points. Hence a's total score is 16. Similarly the sum scores of other alternatives could be determined. This method would give e as the winner as its score is largest (22).

Let us finally consider a comparatively modern method, viz. the approval voting. To make my point I need some additional assumptions which are less plausible than the previous ones, but still show that with this method one *may* end up with yet another winner, viz. d. In approval voting each voter can vote for as many alternatives as he wishes, but for each alternative he can give either 1 vote or no votes. Thus he may not accumulate his votes to any alternative but may only either approve an alternative (i.e.

give it 1 vote) or not approve it (give it no votes). The alternative which has been given the largest number of votes is the winner. Now, assuming that group 1 approves its three highest ranked alternatives, i.e. a, d and e, whereas the other groups approve their two top-most alternatives only, i.e. group 2 b and c and group 3 c and d, the method gives d as the winner because its vote total is largest.

Thus in our example there are five alternatives or candidates, we have used five methods to determine the winners given fixed individual opinions about the order of the alternatives, and we have ended up with a different winner every time. Surely it is not at all a matter of taste which procedure will be adopted by the voting body. The above example is especially constructed so as to point out the differences of methods. The five methods do not always result in different collective decisions. Nor do they exhaust the procedures used in voting bodies. Similar discrepancies in the methods used have been described by Riker (1982, pp. 29-40). The possibility of such phenomena is, however, sufficient to justify the research into the properties of the various voting procedures.

In what follows we shall encounter a number of situations in which different procedures result in different collective decisions. From the view-point of the user of such procedures it is important to know what kind of properties - good and bad - the procedures possess qua collective decision making methods. It will turn out that each procedure is characterized by several shortcomings. If one were to choose a collective decision making procedure one would, of course, be well-advised to take one with as few and/or as unimportant vices as possible. To make such a choice one would naturally need to know which properties characterize any given procedure. This is the topic to be dealt with in the following chapters.

Rather few of us are in a position to actually determine the procedure to be used in a voting body. By studying the properties of a number of procedures which are actually being used or have been put forward for possible use, one can, however, learn which kinds of problems they can lead to. Such a knowledge is sometimes useful in the design of auxiliary devices to overcome the inherent difficulties of the procedures and even when it is not possible to avoid such difficulties, the awareness of them is useful in removing the air of mystique surrounding one of the basic instruments of democracy, viz. voting. Thus, whether one is a designer or user of voting procedures or one's welfare is dependent on the decisions made by collective bodies, it would seem worthwhile to ponder upon the properties the voting procedures have.

CHAPTER 2

PRELIMINARIES

One of the basic methodological dichotomies in the social sciences differentiates methodological individualists from holists. In the theory of voting and representation we encounter approaches that would most naturally be classified as individualistic in the sense that the starting point of the theorizing is the human individual or, more precisely, some specific aspects of human individuals. However, there are grounds for arguing that just the opposite is the case, viz. the theory shows that the collectivities of individuals have properties that clearly indicate the untenability of the individualistic view. Methodological individualism is usually construed as a view according to which all social facts are reducible to those pertaining to individuals. In the following chapters we shall encounter a number of examples pointing to the impossibility of understanding the choices of collectivities in terms of considerations related to individuals only. The additional facts needed to produce satisfactory explanations of the choices of the collectivities are rules or procedures through which the individual opinions are aggregated. For the purpose of introducing the vocabulary of the theory it is, however, sufficient to point out that human individuals are the point of departure of the theory.

We consider the set N of individuals. The set is assumed throughout this book to be finite and the number of individuals in N is n. Another set that is frequently encountered is that of candidates or policy alternatives as the case may be. This set is denoted by X. Most often we shall discuss situations in which X is finite, but some results commented upon later assume that X is a subspace of an n-dimensional real space and, consequently, consists of an infinite number of points (alternatives). Be that as it may, the cases in which X is assumed to be infinite will be clearly indicated. Otherwise we assume that X is finite as well.

The aspect of human individuals we are interested in is their preferences over the set of alternatives. These may reflect the tastes, values or similar considerations that enter a person's deliberations concerning the desirability of alternatives. In accordance with the bulk of social choice theory we assume that the individuals have pretty definite views about the subjective merits of the alternatives. More specifically each voter i has binary relation R_i defined over the set X of alternatives. This relation is called the relation of weak preference. For any a and b in X, a R_i b means that person i considers a at least as desirable as b. This interpretation thus allows for a and b to be considered equally good by i as well as for those cases in which i deems a strictly better than b.

With the aid of the R_i relation we can define the relation P_i of strict preference and I_i of indifference as follows:

a P_i b iff a R_i b and it is not the case that b R_i a,
a I_i b iff a R_i b and b R_i a.

Now R_i so defined is a fairly mild assumption. But the social choice theory usually makes a further assumption, viz. that R_i is
    (i) complete or connected, and
    (ii) transitive.
Together these assumptions imply that the individuals' views of the alternatives are well thought out in the following sense. Completeness means that for any two alternatives a and b in X, person i has either the preference a R_i b or b R_i a or both. In other words, he has actually compared all alternatives with each other. This requirement thus amounts to that for all persons each pair of alternatives is comparable. Transitivity, in turn, means that the fact that a R_i b and b R_i c, implies that a R_i c, for all triples a, b, c of alternatives. Of course, completeness and transitivity do not per se guarantee that each person has a good case for arguing that his preferences are more sensible or rational just because they are complete and transitive. But these conditions imply that a preference order of alternatives can be constructed which is impossible if either of these conditions is not satisfied. These assumptions are quite standard in social choice theory.

Three observations on the relationships between R_i, P_i and I_i are in order. First, we assume that R_i is complete and transitive. Of course, this is a less stringent requirement concerning the individual preferences than assuming that P_i is complete and transitive. Secondly, the fact that R_i is complete and transitive implies that I_i is transitive. Similarly, the same properties of R_i imply that P_i is transitive. The completeness of either I_i or P_i is not implied. Thirdly, the transitivity of R_i implies the transitivity of P_i and the transitivity of I_i. Thus, the transitivity of R_i is not a less stringent requirement than the transitivity of either P_i or I_i, even though the requirement that R_i be both complete and

PRELIMINARIES

transitive is less stringent than the requirement that P_i be both complete and transitive.

Now it is of course a matter of contingency whether individuals in general have complete and transitive preference relations over alternatives. In some cases the assumption seems quite natural and straight-forward, while in others one would feel strained in making such an assumption. Intuitively speaking transitivity fails typically in cases where the criterion by which a and b are compared is not the same as that by which b and c are compared. For instance, I might prefer candidate a to candidate b in local elections because I think a has more sensible views about the traffic plans in the center of my home town. On the other hand, I might prefer candidate b to candidate c because of the former's more appealing personal appearance. But I might still prefer c to a if I thought that his attitude towards homeless people is more in line with my own views. In this case we have more than one "dimension" along which the candidates are compared. In fact what we have here could be viewed as an individualistic extension of a group choice problem (Kelsey, 1984).

CHAPTER 3

SOCIAL WELFARE FUNCTION, SOCIAL CHOICE FUNCTION
AND VOTING PROCEDURES

Given the assumptions concerning the individual preferences over alternatives, the task of social choice theory is to investigate the features of various methods used in aggregating these preferences into collective decisions. These are usually interpreted as collective preferences, although as will be seen shortly there are some problems involved in this interpretation.

There are various ways in which the task of preference aggregation could be approached:

(1) One could look for exactly the same type of preference relation on the collective level as one is assuming on the individual level. In other words, one could take the view that collectivities are no different from individuals and that, consequently, the properties that characterize individual preferences also characterize collective ones.

(2) Quite often in practical decision making one is not interested in anything else but finding the alternative(s) that the collective body "regards" as best. This approach would in principle allow for ties between several alternatives. Indeed, in its extreme form it would even include cases where all the alternatives in X are considered best.

(3) Sometimes there simply is room for no more than one alternative or candidate. In these cases methods otherwise similar to those of (2) could be used but the ties must somehow be broken so that in the end only one alternative is chosen.

The approach (1) is perhaps best-known in social choice literature due to the classic work of Arrow (1963). Arrow tried to find a general method of finding out the collective preference relation R given individual preference relations R_i, so that R would satisfy the same conditions, i.e. completeness and transitivity, as R_i. It turned out that discovering a general method for the construction of such an R would not be impossible, but would become so if a few other innocent-looking conditions would have to be fulfilled as well.

SOCIAL WELFARE FUNCTION

What approach (1) then looks for is a *social welfare function* in the sense of Arrow. It is a function f mapping the n-tuples of complete and transitive individual preference relations into the set of complete and transitive social preference relations. Formally:

$$f: R_1 \times R_2 \times \ldots \times R_n \to R$$

where R_i (i=1, ..., n) is the set of all possible preference relations of individual i satisfying completeness and transitivity. R, in turn, is the set of all possible complete and transitive social preference relations.

Now obviously, one could often do with less than a function f as defined above. In approach (2) one constructs *social choice functions* of the following type:

$$F: X \times R_1 \times \ldots \times R_n \to 2^X$$

where X is the set of alternatives and 2^X is the set of subsets of X. Thus for each set of alternatives and a preference n-tuple or preference profile, F gives a subset of alternatives, viz. the socially "best" alternatives. Approach (2) is adopted in this book. It has been utilized earlier by Fishburn (1973) and Plott (1976) among many others. Actually one could distinguish yet another approach (2') which lies between approaches (2) and (1). (2') aims at ending up with a non-empty choice set in all situations, but, in contradistinction to approach (2), imposes a restriction on the collective preference relation underlying the choices. This additional restriction amounts to the requirement that the collective preference relation be acyclic, i.e. for all sequences of alternatives $x_1, x_2, \ldots, x_k$ such that $x_1 P x_2, x_2 P x_3, \ldots, x_{k-1} P x_k$, it must be the case that $x_1 R x_k$. Here P and R refer to collective strict preference and weak preference, respectively (see Sen, 1970, p. 52).

The approach (1) tries to find out methods that result in social preference orders allowing their users to list the alternatives in the order of social preference from best to worst allowing for ties. Since it may not always be of any consequence for the candidates not chosen which was their position in the social preference order (the essential thing being that they were not chosen), one could think of approach (1) as resulting in something of a luxury. Moreover, should it turn out that plausible methods satisfying the requirements of approach (1) are impossible to come by, one could well think of setting somewhat more modest goals for the research. Approach (2) represents such a revision in goal setting.

Approach (3) is obviously a special case of (2) resulting from the additional restriction on the range of F, viz. that it consists of elements of X only. So for each preference profile and alternative set, F indicates which single alternative is best. Following Gärdenfors (1977) we would call such social choice functions *resolute*. The main motivation for using resolute social choice functions is practical: if one post is to be filled, there is very little one can do with results indicating that the social preference relation is cyclic or that there is a choice set consisting of several applicants. What one needs is a justification for the choice of one applicant before all the others. This is what the resolute choice functions purport to do.

The difference between approaches (1)-(3) can be illustrated by means of a fictitious Eurovision song contest. The methods satisfying the requirements of approach (1) - one of which incidentally is being used in the contest - result in a complete and transitive preference order of the songs allowing possibly for ties. The methods in line with approach (2), in turn, indicate which of the songs is best allowing possibly for ties as well. Finally approach (3) searches for methods which necessarily indicate which (unique) song is best.

What, then, is the relationship between the approaches outlined above and the voting procedures? In this book the latter are considered as realizations of social choice functions. In other words, the basic idea underlying the discussion of specific voting procedures is that each of them corresponds to a social choice function, that is, realizes a given social choice function. The correspondence between choice functions and voting procedures is of the many-to-one type. This means that each procedure realizes one choice function, but several procedures may be realizations of the same choice function.

The relationship between choice functions and voting procedures allows us to analyze the latter in terms of the former. The benefit of this research strategy is that in so doing we are focusing on the bare essentials of the procedures leaving aside many such properties that from the point of view of making good choices are irrelevant.

Our focus is then on approach (2). In many cases of practical interest, one would be pleased to have, if not a complete and transitive social preference relation, at least an idea of which of those alternatives that were not chosen came close to getting chosen. There are procedures that give us this kind of information quite readily while in others this kind of knowledge is not easily obtainable. Therefore, approach (1) also needs some attention. So does approach (3) as practically all voting procedures currently used allow for ties and yet often not more than one alternative is to be chosen. How the ties are broken becomes then an issue of crucial importance.

CHAPTER 4

FIRST PROBLEM: CYCLIC MAJORITIES

Suppose that one is able to construct a procedure that results in a social preference relation that is complete but not transitive. How serious a shortcoming would that be? Very serious indeed, as one could argue that the very meaning of a social choice is in doubt should this happen (see Riker, 1982). More specifically, for a non-transitive social preference the fact that a R b and b R c does not imply that a R c. One important instance of a nontransitivity is the case in which a R b and b R c, but c R a. Such a paradox has been known for centuries and nowadays bears the name of Marquis de Condorcet.

4.1. The Condorcet paradox

Consider a three-member voting body, confronted with the following problem.

Example 4.1. $|N| = 3$, $X = \{a,b,c\}$

The alternatives could e.g. be various energy political options. Thus, a could be the establishment of a new large nuclear power plant, b the establishment of a large coal power plant and c the satisfaction of increased energy needs without large power plants (e.g. by energy saving measures).
Suppose now that the individual preferences are the following:

person 1	person 2	person 3
a	b	c
b	c	a
c	a	b

Thus person 1 regards the establishment of a nuclear power plant as the best alternative and sees the coal plant as the next best one. Person 2, on the other hand, thinks that the possibility

FIRST PROBLEM

of a major accident in a nuclear plant is to be avoided at all cost. Hence, a is his worst alternative. However, person 2 deems the need for additional energy so important that such necessarily limited devices as a saving in energy consumption will not be sufficient. Hence person 2 sees b as the best alternative. Person 3 finally sees c as best alternative because it puts the least additional burden on the environment. Moreover, person 3 views the risks related to the use of nuclear energy as smaller than the damage caused by burning coal. Observe that the preferences are all strict in this example. The alternatives between which a voter is indifferent will, according to the notational convention adopted in this book, be written on the same row.

Suppose that the voting body uses pairwise comparisons and the simple majority rule to construct a social preference relation. Let the agenda of pairwise comparisons be the following:

1. ballot: a versus b
2. ballot: b versus c.

If the voters always vote for the alternative they most prefer among the ones involved in the pairwise comparisons, the result of the first ballot is the victory of a as persons 1 and 3 vote for a. In the second ballot the winner is b with two votes (persons 1 and 2) to one (person 3). The social preference then becomes: a R b and b R c. By transitivity one would then infer that a R c. This is, however, not the result if the pairwise comparison is performed between a and c. By two votes (persons 2 and 3) to one, c wins. Hence if the social preference relation is constructed on the basis of pairwise comparisons with a simple majority deciding the result of each comparison, the result is a violation of transitivity. This phenomenon is known as the Condorcet paradox after the French eighteenth-century mathematician and social philosopher.

The essence of the paradox is that there is no socially best alternative in this situation. No matter which of the alternatives is picked the majority of voters would prefer another alternative to it. There is what is known as a majority preference cycle through all the alternatives: a R b, b R c, c R a. In our example 4.1. the result would mean that the establishment of a nuclear plant is preferred to that of a coal plant by the body, and the establishment of a coal plant would be preferred by the same body to no large plant at all, but the latter would, however, be preferred to the nuclear plant. Thus, no matter which single policy alternative is picked from the set of alternatives, there is always a simple majority of voters preferring another alternative to it: if a nuclear plant were proposed, 2 voters would suggest that no plant at all is better from their point of view; if a coal plant were proposed, 2 voters would prefer a nuclear plant to it; and if no plant at all were proposed, 2 voters would again prefer a coal plant to it.

Now the question arises as to how to deal with cyclic majorities. Obviously, the notion of majority preference in each pairwise comparison would seem unobjectionable. Is there then any way in which the cycles can be resolved without jeopardizing the majority principle and without making the choice essentially arbitrary? In the following we shall first look at a way of "solving" the problem by ignoring it. Thereafter, some results of the theoretical aspects of the problem are discussed.

4.2. How to conceal the problem: the amendment procedure

In the United States Congress as well as in many European legislatures we encounter a voting procedure that is called the amendment procedure. It is based on pairwise comparisons of the alternatives with the majority principle used in each comparison to determine the winner (see e.g. Riker, 1982). Basically it is identical with the procedure outlined in the preceding section. Now, however, we have fixed interpretations attached to the alternatives. For example, consider the following three alternative case. A = "a legislative motion", B = "an amendment to A" and C = "status quo", where a vote for the status quo means that the person in question votes for the rejection of the legislative proposal (either amended or unamended as the case may be).

In the amendment procedure the first ballot is always taken between a motion - i.e. the proposal of a legislative committee - and a proposed amendment. In the case where there is no more than one amendment proposal, the winner of the first ballot is then confronted with the status quo. Finally, the winner of this second ballot is declared the winner of the whole procedure. In the case where there are several amendments to the motion, one of these is confronted with the motion, the winner of this comparison is confronted with the next amendment and so on until all amendments have been taken into comparison. The final comparison is again between the winner of the previous ones and the status quo.

The procedure is clearly based on the assumption that the social preference determined from the pairwise comparisons is complete and transitive. Only by making the latter assumption can one justify the fact that no more than $k-1$ pairwise comparisons are made even though $k(k-1)/2$ is the total number of paired comparisons of k alternatives.

FIRST PROBLEM

Example 4.1. can be used to illustrate the problems related to the amendment procedure. If a and b are a legislative motion and an amendment to it, respectively, while c is the status quo, the first ballot results in the victory of the original motion over its amended form. In the second ballot the winner a of the first one is confronted with the status quo c. In this comparison c wins and is, thus declared the over-all winner. As was pointed out above, however, there is no natural way in which c could be called more preferred by the voting body than any other alternative. In the comparison that was not performed b, the loser of the first ballot, would have beaten c, the over-all winner. Hence, with some justification it could have been chosen as the winner. But then, a beats b and so on.

In short, the way in which the amendment procedure resolves the Condorcet paradox is not very convincing as long as one wants a procedure to single out winners on the basis of some systematic, that is nonarbitrary, criterion. The fact that c wins in Example 4.1. if the amendment procedure is used is due to the special position that the status quo occupies in the procedure. It is always present in the final comparison. The amendment procedure does not satisfy the property called neutrality just because of this unequal treatment of alternatives. A procedure which does not discriminate for or against any alternative is neutral.

On the other hand, the amendment procedure performs in an intuitively plausible way when there is an alternative that would defeat all the other alternatives in a pairwise comparison should the voters vote at each stage according to their preferences. Such an alternative is called the Condorcet winner and we shall encounter it many times later on. Now if there is a Condorcet winner in the set of alternatives and if the voters always vote "sincerely", i.e. in each comparison for that alternative which they prefer, then the amendment procedure results necessarily in the Condorcet winner. Let us consider the following example (see also Riker, 1982).

Example 4.2. $|N| = 3$, $X = \{a,b,c\}$

person 1	person 2	person 3
a	b	c
c	c	a
b	a	b

Observe that persons 2 and 3 have exactly the same preferences as in Example 4.1. and the only change that has occurred is person 1's preference between b and c. Let us interpret the alternatives in the same way as above. Then the amendment procedure calls for the first ballot between a and b. The winner of this ballot is again a, the unamended motion. In the second ballot a is confronted with the status quo c. The result is the victory of c as was the case above. But now there are obvious grounds for arguing that c should be chosen because it not only defeats a, but would also beat b should one make this comparison. The trouble with the amendment procedure is that when the voters vote sincerely and k-1 comparisons are made, one never knows whether the alternative declared the winner actually would defeat all the others, or most of them or just one of them. Clearly the very definition of the Condorcet winner and the assumption of sincere voting at each stage entail that whenever there is a Condorcet winner it will also be chosen by the amendment procedure. However, the procedure does not tell us when the chosen alternative, if it happens to be the status quo, is a "genuine" (i.e. Condorcet) winner or just wins due to the lack of neutrality of the procedure.

4.3. How common are the cycles

It is often thought that the Condorcet paradox and more generally the phenomenon of cyclic majorities is a bizarre special case of not much theoretical importance. After all, Example 4.1. - or the preference profiles obtained from it by relabelling the alternatives or voters - is the only three-person three-alternative profile that generates cyclic majorities, while configurations like in Example 4.2. - or in profiles obtained from it by relabelling the alternatives or voters - are intuitively more common. Can one make any predictions as to how often there are majority cycles?

The first development in this field was the classic work of Black (1958) in the case where only one policy dimension is involved. The voters are represented by points along the one-dimensional continuum. Black showed that as long as the voters have single-peaked preferences over this continuum, there can be no majority cycles. That is, the single-peakedness condition is sufficient to guarantee the absence of cyclic majorities.

Later this condition was generalized by Vickrey and Sen in what is known as the value-restriction on preferences (see Larsson, 1983; Sen, 1966; Vickrey, 1960). Essentially this restriction says that whenever the voters all agree on the position - relative to others - of some alternative, there can be no majority cycles. For example, the single-peakedness condition says that the voters agree on which alternative is not worst.

To illustrate what single-peakedness means let us consider the preference relation of one individual over the set of three alternatives a, b, c. Let us represent the preference relation of the individual over these alternatives in a two-dimensional space so that the points on the horizontal axis represent the alternatives and the points on the vertical axis in decreasing order the ordinal numbers of the alternatives in the individual's preference order. If the preferences are all strict and there are three alternatives, the individual's preference order can be represented by three points in the coordinate space. Drawing line segments connecting the points from left to right along the horizontal axis, we get a curve which may be either

1) monotonically increasing or
2) monotonically decreasing or
3) have a peak in the middle (i.e. the middle alternative is ranked first and two others second and third) or
4) have a cave in the middle (i.e. the middle alternative is ranked third).

The three first-named possibilities represent single-peaked preferences while the fourth does not. Now obviously an individual's preference relation can always be rendered single-peaked by changing the order of alternatives on the horizontal axis. However, the single-peakedness condition pertains to the collection of individual preferences. Such a collection is called the preference profile of the group in question. A preference profile is single-peaked precisely when the alternatives can be arranged consecutively along the horizontal axis in such a fashion that all individual preferences over them are simultaneously of type 1), 2) or 3). The preference profile of Example 4.1., in contrast, is not single-peaked, i.e. there is no way in which a, b and c can be arranged on the horizontal axis so that no individual would have a preference curve of type 4). In Example 4.2. there clearly is an agreement on this, viz. all voters agree that c is not worst (see Figures 4.1. and 4.2.). Consequently, the sufficient condition for the absence of cyclic majorities is fulfilled.

However, when more than one policy dimension is involved things get more complicated. Let us first outline the basic constituents of spatial models. We are given a group of voters whose preferences are represented as follows. Each voter has an ideal point in a m-dimensional real space E^m which corresponds to the outcome that he would regard as best. Moreover, every point in E^m can be proposed as an outcome and be voted upon. The

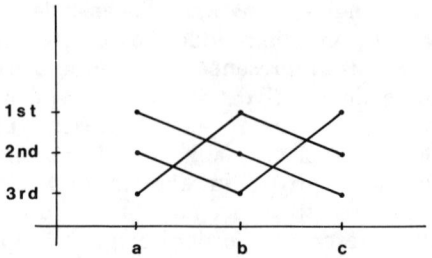

Fig. 4.1. An example of a preference profile which is not single-peaked

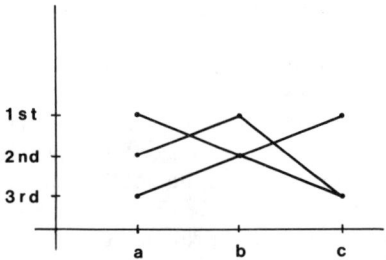

Fig. 4.2. An example of a single-peaked preference profile

preferences of voters over the points in E^m are expressed in terms of the distances of the points from their ideal points. Thus,

$$x \; R_i \; y \text{ iff } d(x-x^i) \leq d\,(y-x^i)$$

where x^i denotes the ideal point of voter i, x and y are two points in E^m, and d is the distance measure. In other words, point x is regarded as at least as good as point y by voter i if and only if x is no further from i's ideal point than y. There are many possible functions one could utilize in measuring distances in E^m. Perhaps the best-known is the Euclidean distance:

$$d(x-y) = ((x_1-y_1)^2 + (x_2-y_2)^2 + \ldots + (x_m-y_m)^2)^{\frac{1}{2}}$$

where x_j and y_j $(j = 1, \ldots, m)$ denote the j'th coordinates of points x and y, respectively.

The individual utilities of alternatives in E^m could then be defined e.g. as follows:

$$U_i(x) = f_i d(x-x^i)$$

where f_i is monotonically decreasing.

Now, in this context Kramer (1973) showed that the condition of single-peakedness, which is sufficient for acyclicity, is stringent when more dimensions are introduced. That is, one sufficient condition for acyclicity is not likely to be fulfilled. However, the single-peakedness is just a sufficient not necessary condition for acyclicity. Obviously, if a majority of voters has the same ideal point in the many-dimensional space, then this point is the Condorcet winner no matter what kind of indifference curves the voters have in the space. But for the avoidance of the Condorcet paradox one does not need such a strong requirement on voter preferences. A less stringent condition is the existence of what is called a *core*. This is a set of alternatives which do not necessarily defeat all the others in pairwise comparisons by a simple majority (as the Condorcet winner does), but which are defeated by no other alternatives by a simple majority. Slightly more formally a point x in E^m belongs to the core (or is a core alternative) if there is no alternative y in E^m such that y is preferred to x by a simple majority.

Now obviously, if there is a core alternative, it cannot be defeated by any alternative per definitionem. Hence, no majority cycle can go through it in E^m. There can be cycles outside the core, but a core alternative cannot enter any such cycle. So even though the presence of a core alternative does not exclude cycles, it at least makes it possible that an alternative that is not defeated by any of the ones in the cycles by a simple majority, can be found. Moreover, no alternative in the cycle satisfies this condition of non-defeatability. One could then expect that whenever a nonempty core exists, some or all alternatives belonging to it will be chosen as soon as a core alternative is put into the voting agenda. So the presence of the core restricts the cycles to (at most) alternatives of less importance.

But it is conceivable that there is no core alternative. This means that for all alternatives x in E^m there is an alternative y in E^m which defeats it by a simple majority. Suppose now that the voters have circular indifference curves in E^m, i.e. they measure the value of each alternative as a monotone decreasing function of its Euclidean distance from their ideal point. An indifference curve of an individual contains all points that are equally preferred by the voter in question. In other words, from his view-point it makes no difference which point on a given indifference curve of his is chosen. It makes a difference, though, on which indifference curve the point lies.

Now if a voter's indifference curves are circular, it means that each circle drawn using his ideal point as the center represents outcomes that are equally good from his viewpoint. A voter with circular indifference curves can be thought of as measuring the value of any alternative by means of the length of the radius of the circle drawn using his ideal point as the center through the point: the longer the radius, the worse the alternative represented by the point.

In this context the absence of the core alternative has quite interesting consequences as has been shown by McKelvey (1976). In particular, a majority trajectory can be build from any alternative x to any other alternative y in E^m, i.e. any alternative can be rendered the majority winner if one can freely introduce new alternatives into the voting agenda and can stop the process at any time and declare the winner of the last pairwise comparison the over-all winner. One must assume that the voters are entirely myopic, that is, they always vote for the alternative which is closer to their ideal point at each stage of the process. Essentially this result means that a majority cycle can be built through any alternative in E^m. McKelvey's (1979) later result shows that this omnipresence of majority cycles holds whenever the voter preferences can be represented as continuous utility functions in E^m (see also Cohen, 1979).

McKelvey's results have interesting theoretical implications. If the agenda-setter knew precisely the ideal points of voters, the circularity of their indifference curves as well as the fact that the core is empty (which he could determine on the basis of his knowledge of the indifference curves), then he could effectively determine which alternative will be chosen by the voting body provided that he can completely decide which agenda is adopted. The agenda-control would thus mean the control of the voting body under the above conditions. The agenda-control must, however, be complete and the voters utterly myopic for the agenda-setter to succeed. If he for example wished to get a point far from the voters' ideal points chosen by the body, it would be more than likely that the voters would smell a rat and propose - if they could - a point nearer to the region surrounded by their ideal points. Similarly, if the agenda were known to the voters even though they had no control over it, it could well occur to them to vote "strategically", i.e. for a point which is actually farther away from their ideal point than the proposed amendment in order to avoid the final outcome planned by the agenda-setter.

As was just pointed out, the method which the agenda-setter can in principle use is based on the fact that a majority cycle can be constructed through all the alternatives. However, these cycles may consist of alternatives very far apart. Consequently, if one has to remain within a restricted subspace of E^m it is not in general possible to construct a majority cycle through every point in the subspace even when there is no core.

The results of Greenberg (1979), Schofield (1984) and Nakamura (1978) show that the number of alternatives in the finite case, the number of dimensions of the policy space, and the so-called Nakamura number determine the possibility of cycles. The result of Kramer (1977), on the other hand, establishes a relationship between the acyclicity and the minmax number. Let us briefly outline the content of these results.

First, Kramer's finding is that if one is willing to require larger than simple majorities in decision making, the collective preference cycles can be avoided (see also Slutsky, 1979). In particular, if in order to get passed a motion needs at least n^+ out of n votes, where n^+ is the minmax number, then the collective preference constructed on the basis of the pairwise comparisons produces an acyclic preference relation. The minmax number n^+ is defined as follows.

For each alternative x, there is obviously one or more alternatives that are its toughest competitors in the sense that the support for them in the pairwise contest with x is larger than the support for any other contestant of x. The support for x may be larger than $n/2$ in all pairwise comparisons which means that x is the Condorcet winner. In these situations the toughest competitor(s) of x get(s) strictly less than $n/2$ votes when confronted with x. But intuitively more typical are situations in which x is defeated by at least one alternative by a simple majority. When defining the minmax number we are looking at the worst showings for all alternatives. Thus, for example, if x defeats all the other alternatives save one, viz. y, by large margins, but is defeated by y with a large margin, its poorest showing - i.e. the defeat to y - is the only thing that counts in the definition of the minmax number. In defining this number we look for the maximum number of votes given to the contestants of any given alternative. We then pick those alternatives for which this maximum number of opposing votes is the smallest or minimum. This minimum is then the minmax number. It is thus typically larger than $n/2$ meaning that all alternatives are defeated by at least one other alternative by a simple majority. But when there is a Condorcet winner the minmax number is less than n/2. More formally, let x be a point in E^m. For any y in E^m let $v(y,x)$ be the number of voters who strictly prefer y to x. Denote by $v(x)$ the maximum of $v(y,x)$ taken over y in E^m, i.e.

$$v(x) = \max_y v(y,x).$$

The minmax number n^+ is now defined as $n^+ = \min_x v(x)$. This result is based on the assumption that the indifference curves of the voters in E^m are circular and that all points in E^m are feasible. The acyclicity referred to in this result concerns the social strict preference relation, i.e. there exists no sequence x_1, $x_2, \ldots, x_t$ of points in E^m such that $x_1 \, P \, x_2$ and $x_2 \, P \, x_3$ and $\ldots x_{t-1} \, P \, x_t$ and $x_t \, P \, x_1$, where each P is determined on the basis of a pairwise comparison with n^+/n-majority rule.

Secondly, the results of Schofield and others relate the existence of majority cycles to the nature of winning coalitions in spatial voting games. A coalition M of voters is winning if and only if it can dictate the social choice from any set of alternatives. A spatial voting game, in turn, is defined by means of the set of alternatives - characterized as in E^m by several dimensions - and the set of winning coalitions. The intuitive background for the results can be seen in Kramer's theorem referred to above. One can thus see that if there is a very influential person in the voting body, i.e. a person whose preference largely determines the social preference, the cycles could be avoided because of the acyclicity of this individual's preference relation. Let us consider the set of all winning coalitions in a voting game. In other words, we consider all those subsets of N that have more than q members if the decision rule is q/n. We denote by D the set of all winning coalitions. One of the constraints which guarantees that there can be no majority cycles can be expressed in terms of the so-called Nakamura number.

The Nakamura number $\bar{n}$ is the minimum cardinality of a subset D' of D which has the property that the intersection of the sets of D' is empty (Schofield, 1983b). In other words, in computing the Nakamura number one needs to know

(1) the set of winning coalitions,
(2) the number of those winning coalitions that have an empty intersection.

The Nakamura number is then obtained as a minimum of the latter. In the case of those games that we have been considering, the determination of the Nakamura number is relatively straightforward. In the case of the simple majority rule, we immediately observe that any two winning coalitions have at least one member in common. On the other hand, adding a third winning coalition can always be done so that its intersection with the intersection of the two previous ones is empty. Thus, the Nakamura number of the simple majority rule voting game is 3. The case in which $|N| = 4$ and $q = 3$ is an exception: in this game the Nakamura number is 4 (Schofield 1983a).

Now Schofield (1983a) shows that with m less than or equal to the Nakamura number minus 2 no cycles can occur if individual preferences are convex. This result does not actually add anything to Greenberg's (1978) earlier result which connects the possibility of the cycles to the dimensionality of the policy space as well as to the value of q, the required majority to make a coalition a winning one. Greenberg shows that as long as the number of dimensions is smaller than $q/n-q$, the convexity of individual preferences implies the convexity of social preferences and, hence, the absence of cycles. The fact that the two results amount to the same can be seen from Schofield's definition of the Nakamura number in games where the winning coalitions are determined solely by their sizes. In these games the Nakamura number is $v(n,q)+2$ where $v(n,q)$ is the greatest integer which is strictly less than $q/n-q$. Obviously, Greenberg's result says exactly what Schofield's does. But the idea of the Nakamura number is not thereby shown to be redundant. As a matter of fact it turns out that this number can be related to yet another parameter of the voting games, viz. the number of alternatives considered. It seems thus a rather general determinant of the acyclicity of social preferences. We shall return to this shortly. It is worth pointing out that in the case of the Condorcet paradox, all these requirements amount to the same, viz. that the number of dimensions be no more than one. In other words, in the 3-person 3-alternative case the restrictions boil down to single-peakedness.

The results thus clearly illustrate the validity of Black's claim that the pairwise comparison of alternatives with the simple majority rule leads to a voting equilibrium and a transitive social preference if the individual preferences can be represented as points along one policy dimension. But on the other hand, they also show that for the simple majority rule this is not necessarily the case in higher dimensions. Indeed, as a corollary of the second part of Schofield's classification theorem on voting games, it can be stated that for $|N|$ = odd (even, respectively) number, if the number of dimensions is larger than or equal to 2 (3), the majority cycles are generic and the core is typically empty.

The above results assume that the voting games are played in policy spaces of a given dimensionality. Perhaps closer to the spirit of our previous discussion are results that make no such assumption. Nakamura (1978), in the study already cited, proved that if the number of alternatives is strictly less than the corresponding Nakamura number, then there necessarily exists a nonempty core for any preference profile over X which consists of acyclic individual preferences. For the simple majority rule, this simply says that if the number of alternatives is less than three, there is always a nonempty core. Certainly this is not an astonishing corollary, but it reinforces the widespread belief that the simple majority rule performs quite well for two alternatives. This belief, in turn, is supported by Arrow's (1963, p. 48) possibility theorem for two alternatives.

An earlier result of Ferejohn and Grether (1974) relates the acyclicity and the number of alternatives to the magnitude of the qualified majority needed to carry a motion in the voting body. These authors show that if q (the required majority) is larger than $n(k-1)/k$, there is a nonempty core and the social preference generated is acyclic. This result is, then, of a similar nature to the one achieved by Kramer mentioned above.

In sum, the above results make it evident that the social preference cycles are more probable when

(1) the majority needed to get a motion passed in the decision making body is small,

(2) when the number of alternatives is large and

(3) when the number of policy dimensions is large, than when these features are not present.

Obvious policy recommendations emanating from the previous results are then the following. If you want to make the appearance of social preference cycles as unlikely as possible, make the Nakamura number of the voting game as large as possible. Obviously for a fixed number of voters there is rather little you can do as $q/n-q$ determines the number. All you can do is to make q as large as you think you can afford without stranding on indecisiveness. The Nakamura number is the crucial determinant of the possibility of social preference cycles because it sets bounds to all the considerations (1)-(3) just mentioned.

4.4. Solutions based on ordinal preferences

The amendment procedure is just one of many procedures used for preference aggregation in collective decision making bodies. There are methods which seem capable of a more rational handling of the problem of cyclic majorities than the amendment procedure. In this section some of these methods are introduced. We start with procedures which can be utilized whenever the ordinal preferences of the voters are available.

4.4.1. Schwartz' procedure

When a Condorcet paradox occurs, there seems to be no grounds for arguing that one of the alternatives in the majority cycle is socially more preferred than others. It would seem then that there is a tie between the alternatives in the cycle. The procedure introduced by Thomas Schwartz (1972) is actually an elaboration and generalization of this intuitive idea. Let us start with the formal definition. Let P be a social strict preference relation over X such that it is not necessarily complete or transitive, but asymmetric. This means that a P b implies that it is not the case that b P a for all a and b in X. Obviously, P can be constructed on the basis of majority pairwise comparisons. Let now $S(X,P) = \{X' \subseteq X | X' \neq \emptyset, \nexists y$ in X-X' such that y P x for some x in X'; for all $X'' \subset X' \exists z$ in X-X'' such that z P x for some x in X''}. Schwartz' idea is to equate the social choices with the union of S(X,P)'s, i.e. $F(X,P) = \cup S(X,P)$.

Schwartz' procedure can be introduced in yet another way. One defines the transitive closure P' of the strict majority preference relation P as follows: x P' y iff there is a set of alternatives $a_1, a_2, \ldots, a_k$ such that $a_1 = x$, $a_k = y$ and a_j P a_{j+1} for all $j = 1, \ldots, k-1$. Secondly, one focuses on the asymmetric part P'_A of P' defined as x P'_A y iff x P' y and not y P' x (Sen, 1986, pp. 1091-1092). The Schwartz' choice set is now

$$F(X,P) = \{x \in X | \nexists y \in X \text{ such that } y P'_A x\}.$$

In Sen's terminology, Schwartz' procedure is called the strong closure maximality procedure, "strong" referring to the underlying strict majority preference relation, "closure" to its transitive closure and "maximality" to the criterion of choosing the maximal elements.

Let us consider this choice function in more detail. What is required for a set of alternatives to be the social choice in Schwartz' sense is that

(i) it must be a subset of X (not necessarily proper),

(ii) no alternative exists outside the choice set that would be socially preferred to any of the alternatives in the choice set,

(iii) the choice set is minimal in the sense that if one took any proper subset of it, then the condition (ii) would not be satisfied with respect to it. One could thus argue that S(X,P) is a minimal choice set that is entirely undefeatable by any alternative outside it.

What then is the relationship between the alternatives in S(X,P) assuming that there are several of them? Obviously, according to condition (ii) no alternative outside S(X,P) can beat any alternative in S(X,P). On the other hand, some alternative in S(X,P) can defeat another one in the set, because if there is one alternative that is not defeated by any alternative in S(X,P) or outside it, then this alternative would certainly satisfy the conditions (i)-(iii) and form one S(X,P) in itself. An interesting special case is of course the Condorcet paradox.

In Example 4.1. the choice set under Schwartz' procedure consists of the entire X, as this is the smallest set satisfying condition (ii). More generally, if there is a so-called top cycle set, i.e. a set of alternatives that form a majority cycle none of which is defeated by any other alternative not belonging to the set, then this set in itself forms a S(X,P).

On the other hand, if there is an alternative that would defeat all the others in pairwise comparisons - i.e. a Condorcet winner - then this and only this alternative would be chosen by Schwartz' procedure. This is because condition (ii) would not be satisfied by any other subset of alternatives and the Condorcet winner would certainly be minimal.

The Schwartz choice function would also choose all the alternatives forming a core, whenever such alternatives exist. The reason is that by definition each of the core alternatives is undefeatable by any other alternatives. Therefore, each of them forms a S(X,P) and thus belongs to the choice set defined as their union.

If the Schwartz' procedure is used, no cyclic majorities can occur simply because the procedure chooses all of them in case the cycle is a top cycle. In some cases the ensuing choice is unmanageably large, but a defender of the procedure could argue that there is no natural way of putting one alternative ahead of another given the preference n-tuple.

4.4.2. Dodgson's procedure

During his career as mathematics lecturer at Christ Church College, Oxford, C. L. Dodgson (alias Lewis Carroll) wrote a number of pamphlets on voting procedures. Many of these are not easily accessible in their original form, but some have been partially reprinted in Black's (1958) book. In these pamphlets Dodgson suggests several voting methods and discusses their merits and flaws.

FIRST PROBLEM

In view of the fact that he actually introduced many voting procedures without as it seems indicating a clear preference between them, it would be somewhat unjustifiable to call any of these proposals *the* Dodgson procedure. All the more so as the opinion seems to be somewhat divided as to which, if any, of the procedures he discusses was his own invention. So, for example, Farquharson (1969) calls a point counting method the Dodgson procedure not noticing that the method had been proposed by an eighteenth century French mathematician Jean-Charles de Borda. Modern literature, however, calls the Dodgson procedure a method that can sometimes be used to overcome the problem of cyclic majorities (see Richelson, 1978; Fishburn, 1977).

The procedure is based on the idea that if there is a Condorcet winner, it must be chosen, but in cases where there is no such winner, an alternative that bears the maximally close resemblance to it should be chosen. Now, of course, much depends on what one means by a "maximally close resemblance". A plausible interpretation of this notion is the following: an alternative that can be rendered the Condorcet winner by the minimum number of preference changes of the voters is the Dodgson winner. Obviously, when there is a Condorcet winner, no changes are needed to make one alternative the Condorcet winner. Hence, the choice is the alternative that would meet the Condorcet winning criterion. However, when no such obvious winner exists, we must specify the way in which the preference changes are counted. A fairly straight-forward way would be to construct a k-by-k matrix of pairwise comparisons of alternatives as follows:

	x_1	x_2	. . . . .	x_k	row minima
x_1	-	n_{12}	. . . . .	n_{1k}	$\underline{n}_1$
x_2	n_{21}	-	. . . . .	n_{2k}	$\underline{n}_2$
.					.
.					.
.					.
x_k	n_{k1}	. . . .	$n_{k(k-1)}$	-	$\underline{n}_k$

In the matrix the entries express how many voters prefer the alternative denoted by the row to that indicated by the column. Thus, n_{ij} is the number of voters preferring x_i to x_j. Now, if for some i, the row minimum $\underline{n}_i$ is larger than n/2, then x_i is obviously the Condorcet winner. If no such alternative exists, we could simply add to each entry of the matrix such integers that would be needed to make the corresponding row minima larger than n/2. For each row one could then sum up the numbers added to the elements in that row. Let us denote this sum by c_i

($i = 1, \ldots, k$). In a sense these sums c_i denote the number of preference changes needed to make the alternative x_i the Condorcet winner. One way of determining the Dodgson winner is, then, to pick that alternative x_i (or those alternatives in the case of a tie) for which c_i is at a minimum. This interpretation of Dodgson's method was utilized by the present author in an earlier context (Nurmi, 1983a).

The method can, however, be given another interpretation even though it is far from certain that Dodgson actually was a firm supporter of either of these interpretations (see Fishburn, 1977). In Fishburn's interpretation the preference changes are counted not in binary fashion as above, but taking into account all the changes that are implied by a change from a P_i b to b P_i a. That is, if a and b are not adjacent alternatives in i's preference order, the change from a P_i b to b P_i a involves other preference changes as well, i.e. those related to third alternatives. Suppose that a person's preference order over three alternatives X={a,b,c} is a P_i b and b P_i c. If we now have to consider the preference change from a P_i c to c P_i a, this change entails two changes: firstly c is pushed ahead of b, and secondly c is pushed ahead of a (or alternatively a is taken down from its top position, first past b, and then past c) (see also Fishburn, 1982).

Now for the evaluation of Dodgson's method it is of some consequence to know which of these two interpretations is adopted. Unless stated otherwise we shall adopt Fishburn's interpretation.

In the case of the Condorcet paradox (Example 4.1.) it is, however, of no consequence which one of these interpretations is used; either way one preference change is needed to make each of the alternatives the Condorcet winner. Hence the choice set is the entire X in both cases. To make a the Condorcet winner, person 2's preference c P_2 a *or* person 3's c P_3 a must be changed into a P_2 c or a P_3 c, respectively. In neither case is there a third alternative involved in the changes. Similarly, one change is needed to make b or c the Condorcet winner with no other changes thereby implied.

It is, however, not the case that Dodgson's method would always result in the same choice as Schwartz' procedure. In other words, Dodgson's method is not equivalent to the strong closure maximality. Consider the following example.

FIRST PROBLEM

Example 4.3. $|N| = 3$, $X=\{a,b,c,d\}$

person 1	person 2	person 3
a	b	c
b	d	a
d	c	d
c	a	b

In this example there is a majority cycle: a P b, b P d, d P c, c P a. As for any proper subset of X, there will always be an alternative not in that subset that beats one alternative in the subset, Schwartz' procedure would specify X as the choice set. Dodgson's method gives {a,b} as the choice set, because

(1) to make a the Condorcet winner, one would need to change either person 2's or person 3's preference between a and c, and

(2) to make b the Condorcet winner one would need to change person 1's preference between a and b.

No changes with respect to third alternatives are called for by these changes. To make c or d the Condorcet winner, on the other hand, one would need at least two changes in preferences. If Fishburn's interpretation is adopted, one needs three changes to make c the Condorcet winner (two if the binary counting is performed), because b and c are not adjacent in any of the preference orders. To make d the Condorcet winner two preference changes are called for: one involving person 1's preference between b and d, and another changing person 3's preference between a and d. In conclusion the Dodgson's choice set in this example consists of a and b.

4.4.3. The maximin method

In section 4.3. we encountered the notion of the minmax number. The method which I shall call the maximin method is based on the idea stemming from Kramer's (1977) article referred to above. With some imagination the method can be traced back to Condorcet's writings, but it is unlikely that the method has been studied systematically before Simpson's (1969) and Kramer's (1977) articles.

For a finite X of relatively small cardinality one can introduce the method with the aid of the by now familiar k-by-k matrix of pairwise comparisons as follows.

	x_1	....	x_k	row minima
x_1	-	n_{12} ...	n_{1k}	$\bar{n}_1$
.				
.				
.				
x_k	n_{k1}	....	-	$\bar{n}_k$

The row minima indicate the minimum number of supports given to the alternative corresponding to the row in all pairwise comparisons. The maximin choice set can now be defined simply as follows:

$$F(X,R) = \{x_i \in X | \ \bar{n}_i = \max_j \bar{n}_j\}.$$

In other words, those alternatives are chosen which in comparison with their toughest competitors fare best. Of course, the toughest competitor may be different for each alternative. The intuitive idea of choosing alternatives satisfying this condition is that by choosing them one can be sure that their comparative performance does not deteriorate under a fixed level indicated by the corresponding row minimum.

The Condorcet paradox is solved by this method in the same fashion as by the previous ones: all the alternatives in the majority cycle of Example 4.1. are chosen. This is due to the fact that the minimum support for each of them is 1. Hence the entire X is chosen. In Example 4.3. the maximin method would choose the entire X again as can be seen from the following matrix of pairwise comparisons:

	a	b	c	d	row minima
a	-	2	1	2	1
b	1	-	2	2	1
c	2	1	-	1	1
d	1	1	2	-	1

Upon looking at the matrix one can immediately see why the maximin choice would sometimes seem unfair: c and d lose two contests and win only one, while a and b lose one contest only. Yet the fact that one contest is lost is sufficient in this example to make the minimum support of the alternatives equal.

In Example 4.3. then Schwartz' method and the maximin procedure result in the same choice set. This is, however, not the case in general. The following example illustrates this.

FIRST PROBLEM

Example 4.4. $|N| = 3$, $X = \{a,b,c,d\}$

person 1	person 2	person 3
a	b	c
b	c	a
d	d	d
c	a	b

The corresponding matrix of pairwise comparisons is the following:

	a	b	c	d	minima
a	-	2	1	2	1
b	1	-	2	2	1
c	2	1	-	2	1
d	1	1	1	-	1

Thus the maximin method would dictate the choice of X, whereas Schwartz' procedure would choose the top cycle set {a,b,c} only. Thus, the maximin method is not equivalent to the strong closure maximality. {a,b,c} would also be the Dodgson choice set, as the reader can easily verify. Thus also the maximin and Dodgson's methods are nonequivalent.

We have then at least three nonequivalent methods of facing the Condorcet paradox. All call for the choice of the entire cycle set and, therefore, one could protest that they are not really solutions at all. On the other hand, given that all the information we have about the voter preferences is the set of transitive and complete individual preferences and that these happen to be distributed so that the Condorcet paradox ensues, there seems to be no nonarbitrary grounds for arguing that one of the alternatives is better than the others. Indeed, one could go as far as to maintain that all procedures that specify a proper subset of X in the Condorcet paradox discriminate either against (or for) alternatives or against (or for) voters. An example of the former is the amendment procedure in which the status quo alternative has a different role than the other alternatives. Let us take a look at yet another way of solving the Condorcet paradox.

4.5. Solution based on scoring function: the Borda count

This method results in the same solution as the previous ones, i.e. calls for the choice of the entire X, but utilizes the voter preferences in a different way than the above methods. The result is consequently typically different from that resulting from the choice functions discussed in the preceding.

The method called the Borda count was originally proposed by Jean-Charles de Borda in the 1770's. Its interesting publication and adoption history has been described by Black (1958) and the proposal itself has been translated and commented upon by DeGrazia (1953) (see also Nurmi, 1983b). Borda was apparently a man of very clear intuition about what a good voting procedure should do. In his memoir he proposes two voting systems to replace the ones that were used at the time. Both of these proposals are probably the results of the same intuitive uneasiness with regard to the procedures then in use as upon closer inspection they boil down to one, i.e. are equivalent (Black, 1958).

Borda's proposal was to utilize not merely the pairwise preferences of the individuals over the alternatives *but also* the information about the positions of the alternatives in individual preference orders in the definition of social choice. Of course, when the individual preferences are complete and transitive, one can easily construct the individual preference orders from the information about pairwise individual preferences. The crux, however, is that the positional information is actually utilized in determining social choices.

In the computation of Borda winners, one usually assigns $k-1$ points to the alternative ranked first by a voter, $k-2$ points to the second ranked alternative and so on. The alternative ranked last thus receives 0 points from the individual in question. The points given to an alternative by all voters are then summed up to obtain the total score of that alternative. The alternative(s) with the largest total score is (are) then chosen. An equivalent way of computing the Borda winner is to add up the rank numbers of alternatives given by the voters, and choose the alternative(s) with the smallest score. In general, when the Borda count is used, an individual's lowest ranking alternative is given a points, the next lowest $a+b$, the next $a+2b$ etc. The choice of the values of a and b does not affect the result as long as a 0 and $b > 0$. As was just stated one usually sets $a=0$ and $b=1$.

In Example 4.1. the computation of the Borda scores would result in a tie between all the alternatives, i.e. each would get three points in total. Hence, the solution of the Condorcet paradox is of the same nature as we observed in the previous section. To see that the Borda count is not equivalent to any of the previously discussed methods, let us take a look at Example 4.5.

Example 4.5. |N| = 11,   X = {a,b,c,d}

5 persons	3 persons	3 persons
d	b	c
a	c	d
c	d	a
b	a	b

Obviously we have here a Condorcet winner, viz. c. Consequently, Schwartz' procedure would choose it as it is obviously the minimal subset of X that satisfies conditions (i) and (ii) mentioned in section 4.4.1.. Dodgson's procedure would also choose c as it clearly needs the minimum number of preference changes to become the Condorcet winner. Moreover, the maximin method would choose c, for its minimum support in all pairwise comparisons is the majority (=6) which obviously cannot be exceeded by any other alternative (see also Nurmi, 1983a). However, the Borda count results in the choice of d as the following calculation shows:

a: 5 x 2 + 3 x 1 = 13
b: 3 x 3 = 9
c: 5 x 1 + 3 x 2 + 3 x 3 = 20
d: 5 x 3 + 3 x 1 + 3 x 2 = 24.

Thus the Borda count does not necessarily choose the Condorcet winner when one exists.

The Borda count is an example of methods based on a scoring function. In this case the function takes the following form:

$$B_i(P) = \Sigma_{x_j \in X} n_{ij}.$$

This formula gives the Borda score of the alternative x_i given the strict preference profile P. Although prima facie it looks rather different from the computational rule outlined above, it can be seen to lead to exactly the same result (see Black, 1958, pp. 158-159; Fishburn, 1977, p. 471; Hansson and Sahlquist, 1976, pp. 196-197). The reason is that counting the number of ranks below any fixed alternative is tantamount to counting the number of alternatives below it, assuming that the preferences are all strict. If one thus counts all the instances in which the given alternative is preferred to some other alternative, one obviously ends up with the Borda score of the alternative in question. An example may be helpful.

Example 4.6. |N| = 11, X = {a,b,c}

We could again think of the alternatives as the energy options of Example 4.1.

5 persons	3 persons	3 persons
a	b	c
b	c	b
c	a	a

Here we have 5 persons convinced about the necessity of having an increased energy supply with the nuclear energy option preferable to the coal plant one. The rest of the voting body is unanimous about the undesirability of a new nuclear plant, but have different views about the other two alternatives. 3 voters acknowledge the need for an additional energy supply and thus put b as their first choice, while the other 3-person group remains unimpressed by the appeals of the industrialists and puts c at the top of its preference list.

Now the Borda scores are obtained in the rank-counting way as follows:

a: $5 \times 2 = 10$
b: $5 \times 1 + 3 \times 2 + 3 \times 1 = 14$
c: $3 \times 1 + 3 \times 2 = 9$

Five persons thus give a 2 points. This means that they prefer a to the two other alternatives. Hence counting five times the two alternatives below a we get the same result; i.e. the Borda score 10 to a. Similarly, e.g. counting three times the one alternative - viz. a - to which c is preferred gives the same result as observing that three persons rank c second.

The most important thing about this scoring function representation is that it clearly shows that the Borda count is - or can be seen to be - based on pairwise comparisons (Young, 1974, p. 45). Interestingly enough, Borda already proposed the above two forms. The pairwise comparison representation is a special case of the general Borda count in the sense that in the former one assumes that a=0 and b=1 in the computation of Borda scores.

4.6. More general majority cycles

The above discussion of the Condorcet paradox and its solutions by various voting procedures has not paid much attention to the

FIRST PROBLEM 35

fact that the paradox is a very special case in the sense that the majority cycle comprises all the three alternatives and the majorities in each pairwise comparison are of equal size. In more general settings one could well have majority cycles of a more general type such as cycles through a small subset of alternatives and/or with majorities of a somewhat different size in various pairwise contests. We have already encountered situations in which a majority cycle is embedded in a larger set of alternatives in Example 4.4. Now one could ask how the above procedures perform in these more general settings, i.e. in generalized Condorcet paradox situations.

The properties of Schwartz' procedure have already been discussed in this kind of setting. It chooses the entire top cycle set if there happens to be one. If on the other hand, there is a majority cycle through alternatives all of which are defeated by a given alternative or set of alternatives, then obviously that cycle will not be chosen by the procedure. In a sense this procedure does not solve the paradox of cyclic majorities: it either chooses several alternatives of the cycle or none of them. One should, however, bear in mind that Schwartz' procedure sets a minimality requirement upon the choice set. Thus, "unnecessary" alternatives in the cycle are excluded from the choice set.

Both Dodgson's method and the maximin procedure can end up with just one of the alternatives forming the cycle. In some cases they may result in two of the cyclic alternatives. The following two examples illustrate these features. The examples are slight modifications of Example 4.4. as can readily be observed.

Example 4.7. $|N| = 7$, $X = \{a,b,c,d\}$

Perhaps it would be appropriate to think of these alternatives as applicants for scholarship and of N as the nominating committee, as in Dodgson's milieu voting procedures were often used in such settings.

2 persons	2 persons	3 persons
a	b	c
b	c	a
d	d	d
c	a	b

The matrix of pairwise comparisons is the following:

	a	b	c	d	row minima	no. of preference changes
a	-	5	2	5	2	2
b	2	-	4	4	2	2
c	5	3	-	5	*3*	*1*
d	2	3	2	-	2	5

The right-most column indicates the number of preference changes needed to make the applicant indicated by the row the Condorcet winner.

We observe that both the maximin and Dodgson's method would result in c which is one of the applicants in the majority cycle going through applicants a, b and c.

Although applicants a and b could call attention to the fact that they are in the same preference cycle as c, the latter applicant would intuitively have a somewhat stronger case for claiming that his choice was, indeed, warranted because he had been ranked first by more voters than a or b.

Continuing the same example with a different voting body, we get the following situation.

Example 4.8. |N| = 5,   X = {a,b,c,d}

2 persons	2 persons	1 person
a	b	c
b	c	a
d	d	d
c	a	b

The corresponding pairwise comparison matrix is:

	a	b	c	d	row minima	no. of preference changes
a	-	3	2	3	2	*1*
b	2	-	4	4	2	*1*
c	3	1	-	3	1	2
d	2	1	2	-	1	4

Hence both a and b will be chosen if either Dodgson's method or the maximin procedure is used.

In more general settings, then, the methods discussed in section 4.4. are solutions to the problem of cyclic majorities. So is

FIRST PROBLEM

the Borda count: in Example 4.7. the Borda winner is c and in Example 4.8. b as the reader can easily verify.

The cyclic majorities are not, however, the only problem facing collective decision making. It could even be argued that they are not the most important one. After all, when a majority cycle occurs it could be maintained that choosing one of the top cycle alternatives could well be justified by the fact that the chosen one is no worse than the ones that were not chosen. If it is not the best this is due to the "nature of things" (i.e. preferences). One could argue that it is not reasonable to blame a method for the failure to dictate a unique choice when there is no obvious reason to put one alternative ahead of the others. A more serious problem - it could be suggested - is that of guaranteeing that the best alternative will be chosen when one obviously exists. Let us now turn to this problem.

CHAPTER 5

SECOND PROBLEM: HOW TO SATISFY THE CONDORCET CRITERIA

5.1. Condorcet criteria

We have already encountered and used the concept of the Condorcet winner. To repeat, it is the alternative that defeats every other alternative by a simple majority. Surely it would be desirable to have a voting procedure that would always guarantee the choice of the Condorcet winner when one exists. Upon closer inspection several questions arise, however, concerning what exactly is meant by this criterion and how plausible it really is.

Firstly, does the criterion say that when the pairwise comparisons are actually performed, the alternative that wins in all of them should be chosen? Or does it mean that when we look at the individual weak preference orderings over the alternatives and assume that each voter would vote for the alternative he regards as strictly the better of the two alternatives confronting each other in the $k(k-1)/2$ pairwise comparisons and would abstain otherwise, then the alternative that would win all the others is chosen. In the latter case, one obviously does not have to perform the pairwise comparisons in order to determine the winner if the individual preference relations are given. If, on the other hand, the former interpretation is adopted, then one obviously needs the data on all pairwise comparisons. The distinction between these two cases pertains to the difference between sincere and insincere voting. If the voting in each pairwise comparison can be assumed to be sincere, then there is no difference between the two ways of determining the Condorcet winner. Unless stated otherwise, we shall conform to the common usage and call the alternative the Condorcet winner that would defeat all the others in pairwise comparisons if the voters voted according to their strict preferences in each pairwise comparison i.e. voted sincerely. Our basis for determining the Condorcet winner is thus the preference profile.

Secondly, the plausibility of the Condorcet winner criterion is not always undisputable. Consider the following example from

SECOND PROBLEM

Fishburn (1973, p. 147) (see also Fishburn, 1977; and Riker, 1982, pp. 82-83).

Example 5.1. $|N| = 5$, $X = \{a,b,c,d,e\}$

person 1	person 2	person 3	person 4	person 5
a	b	e	a	b
b	c	a	b	d
c	e	b	d	c
d	d	c	e	a
e	a	d	c	e

A moment's reflection shows that a is the Condorcet winner. However, when one compares its performance in terms of ranks with that of b, the following result emerges:

	a	b
First ranks	2	2
Second ranks	1	2
Third ranks	0	1
Fourth ranks	1	0
Fifth ranks	1	0

Obviously b's showing is intuitively better than a's, and yet it is a that would have to be chosen if the Condorcet winning criterion is to be honoured. It is perhaps of some interest to notice that b is the Borda winner in this example, showing that, contrary to widely held belief, in cases where there is a discrepancy between the Condorcet and Borda criterion, it may be the latter which is intuitively more plausible.

There are two criteria that are very close to the Condorcet winning criterion (Fishburn, 1973, p. 146), viz. the Condorcet condition and the strong Condorcet condition. The former requires that the choice set is always a subset of the core. The reader will recall our earlier definition of the core: x belongs to the core if and only if there exists no alternative y such that y would defeat x by a simple majority. The strong Condorcet condition, on the other hand, requires that the choice set be identical with the core, i.e. all core alternatives and only they will be chosen. Both of these criteria assume that there is a nonempty core. They say nothing about the situation in which the core is empty. The same is, of course, true of the Condorcet winning criterion.

Obviously, the strong Condorcet condition is the strongest of these three conditions. Whenever an alternative satisfies it, it also satisfies the other two. Similarly, the Condorcet condition is stronger than the Condorcet winning criterion in the sense that whenever an alternative satisfies it, it necessarily satisfies also the Condorcet winning criterion.

The above Condorcet criteria can without much difficulty be traced back to the works of Condorcet. The Condorcet loser criterion, on the other hand, although similar in spirit is undoubtedly of much more recent origin (see e.g. Straffin, 1980). It has also been called the inverse Condorcet criterion by Richelson (n.d.). The criterion requires that an alternative that would be defeated in pairwise comparisons by each of the other alternatives by a simple majority, will not be chosen. Prima facie, this seems plausible enough. Indeed, it would seem that such an alternative can never be a real contestant but will be eliminated early on in the process of choice. It will turn out shortly, however, that especially in one-stage procedures this is by no means necessarily the case.

5.2. Some complete successes

Suppose now that each voter always votes according to his preference ignoring the possible long-term effects of such behaviour. If this is the case, then the amendment procedure obviously satisfies the Condorcet winning criterion. The reason for this is that whenever a Condorcet winner exists, it will necessarily defeat any other alternative in pairwise comparison with a simple majority. Consequently, once it has been put on the agenda of pairwise comparisons, it will per definitionem beat all the rest of the alternatives on the voting agenda. Furthermore, it will be on the agenda because, even though the amendment procedure does not require that all the pairwise comparisons be actually performed, it requires that each alternative must be present in at least one such comparison. Hence, under the stated assumption the Condorcet winner will be chosen (Nurmi, 1983a).

Now what if no Condorcet winner exists? There might still be a nonempty core. This means that there is at least one such alternative that is not defeated by any other alternative and yet it does not defeat all the others. Consequently, there must be at least one pairwise comparison which results in a tie. Whether the amendment procedure satisfies the Condorcet condition or the strong Condorcet condition depends on how the ties are broken. Let us consider the following example.

SECOND PROBLEM 41

Example 5.2. |N| = 4,   X = {a,b,c}

2 persons	1 person	1 person
a	c	b
b	a	c
c	b	a

If now b is a legislative motion and c an amendment to it, then the agenda is:
  (1) b versus c, and
  (2) the winner of (1) versus a (the status quo).
Obviously, a wins if everyone votes on both occasions according to his preference. But suppose that c is the status quo alternative and the agenda is:
  (1) a versus b, and
  (2) the winner of (1) versus c. Then the winner of (1) is a and in (2) the result is a tie between a and c. Clearly a is the core alternative, but its choice depends on
  (i) the agenda, and
  (ii) the tie-breaking procedure.

So one cannot say that the amendment procedure would *necessarily* choose a core alternative when one exists. Indeed, with the latter agenda we notice that {a,c} - the choice set - is not a subset of the core. Thus, the amendment procedure does not satisfy the Condorcet condition and, a fortiori, not the strong Condorcet condition. On the other hand, the amendment procedure obviously satisfies the Condorcet losing criterion as no matter at which stage the Condorcet loser is compared with some other alternative, it by definition is defeated by the latter.

It is immediately apparent that Schwartz' procedure satisfies the Condorcet winning criterion. Obviously, the Condorcet winner is the minimal subset of X satisfying the two conditions of the Schwartz' choice set. Moreover, by definition this procedure chooses all the alternatives belonging to a core whenever one exists, as we observed earlier. It may, however, choose some alternatives not in the core as well (see Fishburn 1977, p. 481). Obviously the Condorcet loser cannot be chosen by Schwartz' procedure when such an alternative exists.

Thus, Schwartz' procedure satisfies the Condorcet winning and losing criterion but fails on the Condorcet conditions. It is thus a nearly complete success story as far as the Condorcet criteria are concerned.

Rather close to a complete success comes a procedure bearing the name of A. H. Copeland. Variations of this procedure are widely used in football and ice-hockey leagues. The idea is simple enough: an alternative (team, say) is given 1 point for each victory and -1 points for each loss. The score of each alternative is the sum of its points over all pairwise comparisons.

More formally:

$$C_i(R,X) = |\{x_j \in X | x_i \, M \, x_j\}| - |\{x_j \in X | x_j \, M \, x_i\}|$$

where $x_j M x_i$ means that the majority of voters prefers x_j to x_i, R is the preference profile, X the set of alternatives and $C_i(R,X)$ the Copeland score of alternative x_i given R and X. Alternatively, $x_i M x_j$ could be interpreted as "the team x_i defeats the team x_j" whereupon the Copeland score would be the difference between the number of victories and losses for each team.

The Copeland choice function is then formally defined as:

$$F(X,R) = \{x_i \in X | \, C_i(R,X) \geq C_j(R,X) \text{ for all } x_j \text{ in } X\}.$$

Thus the alternative(s) with the largest Copeland score is (are) chosen (see Richelson, 1978).

Now obviously this procedure chooses the Condorcet winner whenever one exists because its Copeland score will be k-1 (k-1 victories and no losses). Clearly all other alternatives have at least one loss so that the score of any other alternative cannot exceed k-3. It is also obvious that the Condorcet loser - when one exists - cannot be chosen under Copeland's procedure. Its score would be 1-k, while the score of any other alternative couldn't be less than 3-k.

Assuming that when two alternatives tie, each gets 0 points from that comparison, we can show that Copeland's procedure does not satisfy the strong Condorcet condition. The following example illustrates this.

Example 5.3. $|N| = 4$, $X = \{a,b,c,d\}$

Suppose that to the energy options of Example 4.1. were added the fourth alternative, viz. a major increase in the imports of electricity. Let us denote this option by d giving a, b and c the same content as in Example 4.1. Suppose now that a group of four people, say representatives of major interest groups, is making a collective decision about the issue and that the individual preferences over X are the following.

SECOND PROBLEM

1 person	1 person	1 person	1 person
d	a	c	b
b	d	d	c
a	b	a	a
c	c	b	d

The left-most person is primarily interested in getting the energy demands satisfied as quickly as possible. He figures that increasing the imports produces this result with the shortest time-lag, building a coal plant takes more time and building a nuclear power plant takes longest. For him, however, the option of energy-saving is the worst. The second person agrees with the first on the worst option, but deems the building of a nuclear plant best because in his opinion the plant would allow for a larger amount of national independence than the second best option, energy-import, which, however, would be better than a coal power plant. Thus the second person would not regard the risks related to the use of nuclear energy as very large, while seeing coal-power as much more harmful to the environment. Energy-saving would, however, in his view be the most harmful option, e.g. due to its effects on industrial competitiveness. The third person represents in a fairly straight-forward fashion the environmentalists' interests at least as far as the two upper-most ranks go. Energy-saving is the best option, followed by the energy-import increase (one could perhaps criticize him for having a local emphasis). The two presumably rather intolerable options a and b are ordered by him so that the coal-plant is the worst because he thinks that acid-rain is the most dangerous of the environmental hazards likely to occur. The fourth person, finally, represents the interests of those who strive for a maximum self-sufficiency in this turbulent world (assuming that coal is locally available). The coal plant is the best alternative having the benefit of a continuous local supply enabling the industry to use more energy. However, saving energy would be the next alternative as it allows for a limited reliance on an external energy supply. The nuclear plant would be operative for rather long periods without recourse to foreign sources, assuming that the nuclear fuel is not locally available. Naturally the increase in energy-imports would be the worst option to this person.

The matrix of pairwise comparisons is:

	a	b	c	d	Copeland scores
a	-	2	2	2	0 (core)
b	2	-	3	1	0
c	2	1	-	2	-1
d	2	3	2	-	1 (core & Copeland winner)

Were Copeland's procedure utilized in this decision making situation the winner would be d, the increase in energy-imports, even though both a, the establishment of a nuclear plant, and d belong to the core.

Thus, Copeland's procedure does not guarantee that all alternatives in the core will be chosen. As a matter of fact, it may even turn out that the Copeland winner and the core are distinct alternatives. This is illustrated by the following example.

Example 5.4. |N| = 4, X = {a,b,c,d,e}

This time one could think of a four-member committee choosing from five applicants a persons for a given task. The decision making body could e.g. be the executive board electing the managing director for a company. Let us assume that the preferences of the board members concerning the applicants are the following.

2 persons	1 person	1 person
a	c	d
c	b	b
e	d	e
d	e	a
b	a	c

Obviously applicants a and c are the most controversial ones, a being ranked first by two persons and last by one and c being ranked first and last by one person. Also b is a rather controversial candidate, d being less so. Opinions about e, on the other hand, are rather uniform.

The corresponding matrix of pairwise comparisons is:

	a	b	c	d	e	Copeland scores
a	-	2	3	2	2	1 (core)
b	2	-	1	1	2	-2
c	1	3	-	3	3	2 (Copeland winner)
d	2	3	1	-	2	0
e	2	2	1	2	-	-1

Candidate a is defeated by no other alternative and belongs thus to the core being the only candidate with this property. Candidate c, on the other hand, is the Copeland winner, but is defeated by one candidate (viz. a). Hence in this example the Copeland winner does not belong to the core. In this example then the applicant a who was not elected for the job might well feel that he has not

SECOND PROBLEM

been treated fairly. After all he beats the chosen alternative c in a pairwise contest, while no alternative could beat a. As long as the board is committed to Copeland's procedure and keeps its preferences fixed, a's complaint is of no avail.

We already observed that the Borda count does not always choose the Condorcet winner when one exists. Undoubtedly the first to point this out was the Marquis de Condorcet who argued that no matter which non-negative points are given to the alternatives, the Borda count cannot guarantee the choice of the Condorcet winner as long as the most-preferred alternative is given more points than the next-most preferred and the latter in turn more points than the third alternative etc. (Black, 1958, p. 179).

What Condorcet did not point out is that the Borda count never chooses a Condorcet loser. Indeed, it can be argued that this was the main motivation for Borda when he introduced his new system (Richelson n.d.; Ray private communication). The reason why the Condorcet loser is never chosen by the Borda count can be presented as follows.

Let an alternative a be the Condorcet loser. Then it must be the case that in each pairwise comparison it gets less than $n/2$ votes. Thus its Borda score is strictly less than $(n/2)(k-1)$. We can now show that if every alternative gets this same score, the sum of the scores will be strictly less than the total number of scores summed over all the alternatives, i.e. $\Sigma_{x_i \in X} B_i(P)$.

The latter quantity, i.e. the total number of scores can simply be obtained as $n(k-1)+n(k-2)+ \ldots n(k-(k-1))$, because every voter gives some alternative k-1 points, some alternative k-2 points etc. Now if every alternative gets the same score as the Condorcet loser, the total sum of scores will be less than $(n/2)k(k-1)$, or $(n/2)(k^2-k)$. It is a simple exercise in mathematical induction to show that this quantity is identical with $n(k-1)+ \ldots +n(k-(k-1))$, whenever $k \geq 2$. Obviously the number of voters has no effect on the relationship between these quantities because it features as a multiplier in both expressions. Hence we have to compare $(k^2-k)/2$ with $k(k-1) - (1+2 \ldots +(k-1))$. For $k=2$ both expressions have the value 1. Suppose now that the expressions are identical for $k=t$. We shall show that they are then equal for $k=t+1$ as well. For the latter expression we obtain: $(t-1)t + 2t - (1+2+ \ldots +t)$ or $t + (t-1)t - (1+2 \ldots +(t-1))$, thus when the value of k is increased from t to t+1 the value of the expression increases by t. For the former expression we get $(t+1)t/2 = ((t-1)t+2t)/2$. Consequently, the value of the former expression has increased by t as well. Thus the two expressions are equal for $k=t+1$. Hence they are equal for all values of $k \geq 2$.

Now as was pointed out above the sum of the Borda scores, assuming that they all equal the score of the Condorcet loser, is strictly less than one of the expressions we just showed to be equal. In other words, if each alternative had the Borda score of a Condorcet loser, then the sum of the scores would be less than required. Therefore, some of the alternatives must have a strictly larger Borda score than the Condorcet loser. Hence, the Condorcet loser cannot be the Borda winner.

The fact that the Borda winner and the Condorcet winner may not coincide, of course, implies that the Borda count does not satisfy either the strong Condorcet or Condorcet condition. However, by an ingenious modification Nanson constructed a method which, although based on the Borda count in every phase, yet guarantees the choice of a Condorcet winner whenever one exists. Nanson's method is a multi-stage one. First the Borda count is applied to the entire set X of alternatives to determine which alternative has the lowest Borda score. This alternative is then eliminated from further consideration. The Borda scores are then recomputed for the rest of the alternatives and again the alternative (or in the case of a tie, the alternatives) with the lowest score is (are) eliminated. The procedure is continued until only one or, in the case of a tie, several alternatives with the same Borda score are left.

How does this method guarantee the choice of an eventual Condorcet winner? The reason is as follows. If an alternative, a say, is the Condorcet winner, then its score must be strictly larger than $(n/2)(k-1)$ because it defeats all the other alternatives ($k-1$ in number) by a simple majority. If every alternative gets a score of more than $(n/2)(k-1)$, then by a similar reasoning as above we can conclude that the sum of the Borda scores of alternatives is larger than the total sum of the Borda scores which is contradictory. Consequently, at least one alternative must have a Borda score strictly less than that of the Condorcet winner. This, in turn, means that at each stage the alternative which is eliminated according to Nanson's method is not the Condorcet winner. Therefore, as the method chooses the non-eliminated alternative, it must be the Condorcet winner.

Suppose now that there is no Condorcet winner, but a non-empty core exists. How does Nanson's method behave in such a situation? By a similar argument as above, it can be shown that the core alternatives never get a Borda score less than $(n/2)(k-1)$ and, consequently, are not eliminated. However, some alternatives not in the core may eventually be chosen as well (see Fishburn 1977, p. 481). Thus, Nanson's method satisfies neither of the Condorcet conditions.

SECOND PROBLEM

Another modification of the Borda count is Black's procedure (see Black, 1958). It simply chooses the Condorcet winner when one exists. Otherwise, it chooses the Borda winner. By definition this procedure satisfies the Condorcet winning criterion. It also satisfies the Condorcet losing criterion because whenever there are both the Condorcet winner and the Condorcet loser, it chooses the Condorcet winner, thereby excluding the Condorcet loser. On the other hand, when there is a Condorcet loser but no Condorcet winner, Black's procedure chooses the Borda winner which, as we have just seen, never coincides with the Condorcet loser. Suppose now that there is no Condorcet winner, but yet a nonempty core. Then it turns out that Black's method does not necessarily choose the core alternatives. Consider the following slight modification of Example 5.1.

Example 5.1.' $|N| = 4$, $X = \{a,b,c,d,e\}$

person 1	person 2	person 3	person 4
a	b	e	a
b	c	a	b
c	e	b	d
d	d	c	e
e	a	d	c

Now the only alternative that is not defeated by any other is a. Yet it is not the Condorcet winner because it does not beat e. a is, thus, the core alternative. Black's procedure chooses - in the absence of the Condorcet winner - the Borda winner which is b. Thus Black's procedure satisfies neither the Condorcet nor the strong Condorcet condition.

Most of the above procedures can be viewed as successes as far as the Condorcet criteria are concerned. They all - except the Borda count - guarantee the choice of the Condorcet candidate, i.e. the Condorcet winner, whenever one exists. Moreover, they never choose a Condorcet loser. Let us now turn to some other methods that only partially succeed in terms of these criteria.

5.3. Some partial successes

The Borda count is by no means alone in the group of voting procedures that do not satisfy the Condorcet winning criterion. Indeed, there are two widely used procedures that have the same drawback. In fact one of them has the questionable distinction of not satisfying either of the Condorcet criteria, i.e. failing to choose the Condorcet winner when one exists and possibly choosing the Condorcet loser when it exists. The latter procedure is the plurality voting, one of the most commonly adopted voting procedures in modern times.

The plurality procedure does not necessarily choose a Condorcet winner assuming again that each voter votes according to his true preferences. In the case of plurality voting this assumption implies that the voters vote for that particular alternative they regard as best or most-preferred. Now Borda showed some two centuries ago that plurality voting is not in all respects a satisfactory procedure. Let us briefly look at what is nowadays called the Borda paradox (see DeGrazia, 1953; Colman and Pountney, 1978).

Example 5.5. $|N| = 21$, $X = \{a,b,c\}$

For the sake of illustration one could think of a one-member constituency in British parliamentary elections. Three parties put forward candidates in this constituency, a being the conservative candidate, b the Labour candidate and c the SDP candidate. The reader worried about the size of the constituency can multiply the numbers in this example by, say, 1000 or 10.000 or 100.000 to get a more realistic picture.

7 voters	7 voters	6 voters	1 voter
a	b	c	a
c	c	b	b
b	a	a	c

Now plurality voting gives the following result:

a gets 8 votes
b gets 7 votes
c gets 6 votes

Thus a is the plurality winner. But it is obviously not the Condorcet winner as c would defeat it in pairwise comparison by 13 votes to 8. As c would also beat b by 13 to 8, the Condorcet winner exists and is c, not a. This shows then that plurality voting does not satisfy the Condorcet winning criterion.

Borda's point was, however, to show an even more dramatic failure in plurality voting. In Example 5.5. the plurality winner a is defeated not only by the Condorcet winner c, but also by b. In other words, the plurality winner is defeated by all the other alternatives in pairwise comparisons by a simple majority. This, of

SECOND PROBLEM

course, makes a the Condorcet loser (Borda didn't use this term which is of fairly recent origin). Consequently, Example 5.5. also shows that the plurality procedure does not satisfy the Condorcet loser criterion. In conclusion then plurality voting fails on all our Condorcet criteria, i.e. the winner and loser criterion plus the Condorcet and strong Condorcet conditions.

It is easy to see why there is such a discrepancy between the plurality and Condorcet criteria: the former utilizes the first preferences of the voters only, whereas the latter are based on pairwise comparisons of all alternatives including those at lower levels in the individuals' preferences. Plurality voting is thus an example of *positional voting procedures*, that is, procedures in which the positions of the alternatives in individual preference orders are essential determinants of the social choice. We have already encountered another - perhaps even more illustrative - example of positional procedures, viz. the Borda count. It was Borda's proposal for resolving the above paradoxical situation. As we already observed, it is, indeed, successful in avoiding the choice of a Condorcet loser, but does not guarantee the choice of the Condorcet winner.

Another procedure that is widely used in contemporary elections is the plurality runoff method. In this procedure the first ballot is conducted in a similar manner as in the plurality voting. If one of the alternatives receives more than 50 % of the votes, it will be declared the winner. If no such alternative exists, the second ballot is conducted between those two alternatives that in the first round received the largest plurality of votes. The alternative getting most votes on the second round is then the winner. Now if the Borda count can be regarded as an attempt to avoid the choice of a Condorcet loser, the plurality runoff procedure can also be given the same motivation. It is thus obvious that if there is a Condorcet loser, it cannot possibly be chosen by the plurality runoff method as it will necessarily be defeated in the second round by its contestant (if it manages to get to the second round) because the Condorcet loser is by definition defeated by all the other alternatives including its opponent in the second round. In case there is no second round, it must be the case that the winning alternative is the Condorcet winner because it is ranked first by more than 50 % of the voters assuming again that the voting is strictly according to preferences.

So, the plurality runoff does better than the plurality voting in terms of the Condorcet losing criterion. However, it does no better than the latter in terms of the Condorcet winning criterion. Example 5.5. shows that it can fail to choose a Condorcet winner when one exists. In that example no alternative gets more than

50 % of the first ranks. Therefore, the runoff contest takes place between the two maximal vote-getters, i.e. a and b. As we observed already, neither of these is the Condorcet winner which happens to be c. Subsequently, b defeats a in the second round by 13 votes to 8. Thus, the Condorcet winner is not necessarily chosen even though there is one. The reason for this is apparent: the Condorcet winner is eliminated in the first round where the procedure is essentially plurality voting. It is equally clear that if the Condorcet winner makes its way to the runoff ballot, then it cannot be defeated because in the second round the contest is pairwise or binary and the Condorcet winner by definition beats all its contestants in binary comparisons.

The failure of the plurality runoff method to choose the Condorcet winner can only occur under certain preference profiles. In the three-alternative case the bounds of the relative frequencies of various logically possible preference orders that are conducive to the exclusion of the Condorcet winner, can easily be computed.

Let us first observe that, regardless of the number of alternatives considered, the Condorcet winner will necessarily be chosen by the plurality runoff method if it is present in the second round, i.e. in the runoff. This is because it by definition beats all the other alternatives and in particular that alternative it is confronted with in the second round. Hence if it is in the second round, it will win. On the other hand, for any number of alternatives no alternative can win unless it is present in the second round, provided that the runoff is needed. Thus we see that in cases where there is a second round, those preference profiles that exclude the Condorcet winner in the first round will automatically exclude it from the choice set of the plurality runoff procedure. Let us illustrate this reasoning in the three-alternative case. For X = {a,b,c} we have 6 logically possible preference orders:

1	2	3	4	5	6
a	a	b	b	c	c
b	c	a	c	a	b
c	b	c	a	b	a

Let now n_i (i = 1, ..., 6) be the number of voters having the i'th preference order. Assuming that a is the Condorcet winner, we get:

(1) $\quad n_1+n_2+n_5 > n_3+n_4+n_6$

SECOND PROBLEM

because a defeats b in pairwise comparison. Similarly we get:

(2) $\quad n_1+n_2+n_3 > n_4+n_5+n_6$

because a also defeats c in pairwise comparison.

Now a necessary and sufficient condition for the exclusion of a by the plurality runoff method is that both b and c get more votes in the first round. In other words, it is both necessary and sufficient for a's not being chosen that

$$v_a < v_b \text{ and } v_a < v_c$$

where v_j (j=a,b,c) denotes the number of votes given to j in the first round. As we have assumed that the voters vote strictly according to their preferences at each stage, the above two conditions can be rewritten as:

(3) $\quad n_1+n_2 < n_3+n_4$ and

(4) $\quad n_1+n_2 < n_5+n_6$.

Combining conditions (1), (2), (3) and (4) we get the following two conditions for the exclusion of the Condorcet winner a:

(5) $\quad n_5-n_6 > (n_3+n_4) - (n_1+n_2) > 0$

(6) $\quad n_3-n_4 > (n_5+n_6) - (n_1+n_2) > 0$.

Let us illustrate this with an example.

Example 5.6. $|N| = 5$, $X = \{a,b,c\}$

1 voter	2 voters	2 voters
a	b	c
b	a	a
c	c	b

Clearly a is the Condorcet winner. Now we have $n_1=1$, $n_3=2$ and $n_5=2$ while $n_2=n_4=n_6=0$. Clearly now $n_5 > n_3-n_1 > 0$ and $n_3 > n_5-n_1 > 0$ whereby (5) and (6) are satisfied.

If one wishes to get an idea of how often the exclusion of the Condorcet winner might occur in the real world, one could try to figure out how often there are situations in which two main contestants are strongly disliked by their supporters while there is a third candidate having some support but less than the other two.

The third contestant might well be the Condorcet winner if it is ranked second by the supporters of the two strong candidates and if none of the candidates has more than 50 % of the first ranks. Of course in cases where there are more than three candidates the exclusion of the Condorcet winner can happen as well under the plurality runoff procedure. The conditions conducive to such an exclusion are that the two strongest candidates are controversial in the sense that they are either ranked first or relatively low in the individual preference rankings.

Because the plurality runoff method does not necessarily choose the Condorcet winner when such an alternative exists, it follows that it does not necessarily choose the core alternatives. Thus it does not satisfy either the Condorcet or strong Condorcet condition.

Dodgson's method can also be considered a partial success in terms of the Condorcet criteria. It obviously satisfies the Condorcet winning criterion as the Condorcet winner needs no preference changes to become one, while all the others need at least some preference changes. It does not, however, satisfy the Condorcet losing criterion as can be seen from the following.

Example 5.7. $|N| = 15$, $X = \{a,b,c,d\}$

5 voters	3 voters	2 voters	5 voters
a	b	d	c
b	c	b	d
d	a	c	a
c	d	a	b

The corresponding binary comparison matrix is:

	a	b	c	d	no. of preference changes
a	-	10	5	8	3
b	5	-	10	8	3
c	10	5	-	8	3
d	7	7	7	-	3

Thus Dodgson's method chooses the whole alternative set. And yet there is a Condorcet loser, viz. d. Consequently, Dodgson's method is incompatible with the Condorcet loser criterion.

It can also be observed that Dodgson's method does not necessarily choose a core alternative when one exists. Consider again Example 5.4. The reader can easily verify that the number of preference changes needed to make c the Condorcet winner is 2, while all the other alternatives - including the core a - need

SECOND PROBLEM

more preference changes to become the Condorcet winner. This conclusion is independent of the way in which the changes are counted (cf. above 4.4.2.). It follows then that Dodgson's method does not satisfy either the Condorcet or - a fortiori - the strong Condorcet condition.

Another partial success is the maximin method which necessarily chooses a Condorcet winner when one exists. This is due to the fact that in the pairwise comparison matrix the row minimum of the Condorcet winning alternative is strictly larger than $n/2$. Obviously all the remaining alternatives have at least one entry smaller than $n/2$ in their respective rows. Hence, the choice of the Condorcet winner is guaranteed. However, the exclusion of the possible Condorcet loser is not necessary under the maximin method. The following example shows this.

Example 5.8. $\quad |N| = 3, X = \{a,b,c,d\}$

1 voter	1 voter	1 voter
a	c	b
d	d	c
b	b	d
c	a	a

The corresponding binary comparison matrix is:

	a	b	c	d	row minima
a	-	1	1	1	1
b	2	-	2	1	1
c	2	1	-	2	1
d	2	2	1	-	1

Thus, the entire alternative set X is chosen. However, there is a Condorcet loser, viz. a. Hence the method does not satisfy the Condorcet loser criterion. An even more dramatic demonstration of this is Example 5.7. where the row minima are the following: a 5, b 5, c 5 and d 7. Hence the maximin winner is d which, however, is the Condorcet loser as was pointed out above.

In contradistinction to Dodgson's method the maximin method satisfies both the Condorcet condition and the strong Condorcet condition. The reason is that the minimum entry in each row corresponding to a core alternative is $n/2$. Clearly no other alternative can have a larger minimum support. Thus all the alternatives in the core are chosen by the maximin method. This result assumes that there are no ties in individual preferences.

An interesting procedure designed to guarantee proportionality in representation was proposed by Thomas Hare (1861) in the nineteenth century. It takes on many variations and names, but its most common forms are called the single transferable vote system and the preferential voting or the alternative vote system (Doron, 1979; Merrill, 1984). The former version is utilized in multiple member constituency systems, whereas the latter two denote a special case of the former and are used in cases where one alternative or candidate is to be elected. We shall here concentrate on this special case. The method works as follows. The voters are asked to reveal their preferences concerning the candidates or alternatives. If among the individual preference ballots more than 50 % have a given alternative ranked first, then that alternative is chosen. If no such alternative exists, the alternative with fewest first ranks is eliminated and the rest of the alternatives are pushed upwards in the preference lists of the voters so that those voters who had the eliminated alternative as their first preference will now have their original second preference as their first ranked alternative. Again one determines whether there is an alternative that is ranked first by more than 50 % of the voters according to these new preference orders. If such an alternative exists, it is declared the winner. Otherwise one continues eliminating the alternatives in a similar fashion as above. Eventually one must end up with a winner or a tie when proceeding in this manner.

Now this method does not satisfy the Condorcet winning criterion. As it is in general the case that the failure to satisfy this criterion entails an incompatibility with both of the Condorcet conditions, this is of course also true of the Hare system. On the other hand, the system is compatible with the Condorcet losing criterion, i.e. it never chooses a Condorcet loser. This is due to the fact that in order to become the Hare winner an alternative must have at least 50 % of the first ranks in some subset of alternatives. This is a necessary though not sufficient condition for becoming the Hare winner. Now the Condorcet loser cannot have 50 % of the first ranks in any subset, because if it had, it would defeat all the other alternatives in that particular subset by a simple majority. This, in turn, would mean that it is not a Condorcet loser after all. So, the Hare system necessarily excludes the Condorcet loser when one exists. To see that it may fail to choose the Condorcet winner, consider the following example from Straffin (1980, pp. 23-25).

Example 5.9. $|N| = 17$, $X = \{a,b,c,d,e\}$

SECOND PROBLEM

5 voters	2 voters	3 voters	3 voters	4 voters
a	b	c	d	e
b	c	b	b	b
c	d	d	c	c
d	e	e	e	d
e	a	a	a	a

No alternative has more than 50 % of the first ranks. Therefore, one must resort to elimination. Obviously, b is eliminated first, as it has the fewest first ranks (2). We need go no further, as b is the Condorcet winner in this example. Therefore, the Hare system is incompatible with the Condorcet winning criterion. Again it is easy to see why the Condorcet winner is not chosen: the criterion used in the elimination of alternatives is a positional one, while the Condorcet criterion is essentially binary.

Coombs suggests a minor modification to the Hare system which, however, does not change the incompatibility of the system with the Condorcet winner criterion. Coombs' method is otherwise identical with the Hare system, but the elimination criterion is different: the alternative which is ranked last by the largest number of voters is eliminated first. The winning criterion is the same as in the Hare system: the alternative with more than 50 % of the first ranks is the winner. It is somewhat more difficult to construct an example showing the failure of the Coombs' system to choose the Condorcet winner, but, as was just stated, the system is incompatible with the Condorcet winning criterion. We shall again borrow an example from Straffin (1980, p. 26):

Example 5.10. $|N| = 21$, $X = \{a,b,c\}$

5 voters	4 voters	2 voters	4 voters	2 voters	4 voters
a	a	b	b	c	c
b	c	a	c	a	b
c	b	c	a	b	a

None of the alternatives has more than 50 % of the first ranks. Hence, a is eliminated as it has the largest number of third ranks. However, a is the Condorcet winner as can readily be observed. Coombs' method satisfies the Condorcet loser criterion. The reason is the same as in Hare's system.

Thus both the Hare system and Coombs' procedure are incompatible with the Condorcet winning criterion. It obviously follows that they are also incompatible with the Condorcet conditions. Borda's and Dodgson's methods along with the maximin and plurality runoff procedures satisfy one Condorcet criterion, but not both of them. With respect to the Condorcet conditions we have some variation: the maximin method satisfies both the Condorcet and the strong Condorcet condition; the plurality runoff and Dodgson's method satisfy neither of these conditions. In the case of the plurality runoff the reason is that the method does not satisfy the Condorcet winning criterion, whereas Dodgson's method fails on both Condorcet conditions in spite of its compatibility with the Condorcet winning criterion. We already observed that the plurality procedure does not satisfy either the Condorcet winning or losing criterion. In this respect it is a complete failure as far as these criteria are concerned. Its discussion in the present section is therefore not strictly correct. I have, however, chosen to deal with it here because of its intimate relationship with a procedure that is a partial success, viz. the plurality runoff. The plurality procedure is, however, not the only complete failure with respect to the Condorcet criteria and conditions.

5.4. Complete failures

A relative newcomer in the field of voting systems is approval voting. In the late 1970's it was invented independently by several authors (see Brams and Fishburn, 1983a, pp. xi-xiv). The system was brought to a wider political science audience by an article by its chief proponents, Steven J. Brams and Peter C. Fishburn (1978). The idea behind approval voting is simple enough: instead of asking people to indicate no more than their most-preferred alternative, one should ask them to list all the alternatives that they consider acceptable. However, since it may be difficult for most voters to give a preference order at least if the number of alternatives is large, it is more reasonable to ask them to indicate just the acceptable alternatives without ordering them. Hence the question asked of the voters at the polls is: which of these candidates or alternatives do you consider acceptable with respect to the office or policy at hand? It is, of course, essentially different from the following question: who in your opinion is the best candidate for the office or the best policy alternative? The voters can vote for as many alternatives as they wish in approval voting. In a sense, the system can be characterized as the "one man, k votes" -method (Brams and Fishburn, 1983b, p. 31). Each voter has at his disposal k votes if k is the number of alternatives or candidates. He can give each alternative one vote or no vote. The alternative with the largest total number of votes (approvals) is then declared the winner. We immediately observe that giving k votes is equivalent to giving 0 votes as voting for all candidates

SECOND PROBLEM

or voting for none leaves the differences in vote totals unaffected.

Now this system has been introduced mainly as an alternative to plurality voting. However in terms of the Condorcet criteria, its performance is not, prima facie, any better. Approval voting does not guarantee the choice of the Condorcet winner when one exists. Worse, it can even choose a Condorcet loser when there is one. The following example illustrates these features (see also Nurmi, 1983a, p. 186).

Example 5.11. $|N| = 9$, $X = \{a,b,c\}$

4 voters	3 voters	2 voters
a	b	c
b	c	b
c	a	a

If the voters all approve of their first preference only, the result is the victory of a by 4 votes against 3 and 2. However, a is the Condorcet loser. Moreover, there is a Condorcet winner in this example, viz. b. Thus, approval voting fails on both Condorcet criteria. It follows that it also fails on the Condorcet conditions. In Example 5.11. the same result would follow if one or more voters in groups consisting of 3 or 2 voters voted for all (or none) of the alternatives.

So, there does not seem to be any difference between plurality voting and approval voting as far as the Condorcet criteria are concerned. Upon closer inspection, however, there are some differences. Fishburn and Brams (1981) point out that there is always a collection of sincere voting strategies in approval voting that results in the choice of the Condorcet winner when one exists. By a voting strategy one simply means a subset of X (the alternative set) indicating which candidates the voter gives his approval vote to. A voting strategy S is sincere if and only if $x_i \in S$ implies that $x_j \in S$ whenever x_j is preferred to x_i by the voter in question. This result is of the nature of an existence theorem; it states that it is possible to find such a collection of sincere strategies. So, in Example 5.11. if all voters give an approval vote to the two most preferred alternatives, the voting strategies are sincere and the result is the victory of b, the Condorcet winner. The result does *not* state that all collections of sincere strategies would result in the choice of the Condorcet winner when one exists. In fact, Example 5.11. shows that there are also sincere strategy collections that do not result in the choice of the Condorcet winner. All the voting strategies mentioned in the context of that example are sincere. It is, however, noteworthy that plurality voting does not have this property. In a sense this is quite natural, since there is only one collection of sincere voting strategies in plurality voting and if this does not result in the choice of the Condorcet winner, then obviously the existence result concerning approval voting cannot hold for plurality voting.

5.5. Some probability considerations and the plausibility of the Condorcet criteria

Anyone who has tried to construct examples showing that a given procedure fails to satisfy some specific systematic property, knows that such a construction is for some systems and properties much more difficult than it is when other properties and/or systems are concerned. To be specific, suppose that it is intuitively very difficult to construct an example in which a given procedure does not result in the choice of the Condorcet winner assuming, as we have done in the preceding, that the voters reveal their true preferences. Suppose, moreover, that for some other procedure the construction of such an example under similar assumptions is very easy. Would it, then, be plausible to put the systems into the same class because of their incompatibility with the Condorcet winning criterion?

Intuitively it would seem that compatibility is a property capable of taking on degrees rather than a simple dichotomy. On the other hand, it could be argued that what one is interested in when studying the procedures is their properties as general methods of preference aggregation. One makes thereby no assumption about the frequencies of the various individual preference distributions. Either way, it is obvious that the probability of incompatibility with various criteria provides information concerning the procedures that is additional to the mere statement that the incompatibility exists. It is, however, equally obvious that without any specification of the probability of the relevant individual preference order distributions, no probability statements of this sort can be made.

Some results concerning the probability that various voting procedures violate the Condorcet winning criterion have been obtained from simulation studies (see e.g. Chamberlin and Cohen, 1978; Merrill, 1984). Merrill generates random societies, that is, uniform and independent utility distributions over the sets of candidates. For a fixed number k of candidates he simulates 10 000 elections, each involving 25 voters, and determines the relative frequency with which the various procedures produce

SECOND PROBLEM

results coinciding with the Condorcet winner. Merrill's results are shown in Table 5.1.

	No. of candidates					
Voting system	2	3	4	5	7	10
plurality	100	84.4	76.6	69.7	61.0	51.7
Borda	100	94.1	90.4	88.7	86.2	86.4
approval	100	82.4	77.1	75.3	73.6	70.7
Hare	100	96.4	92.1	88.9	83.3	75.8
pl. runoff	100	96.4	89.4	83.7	71.8	60.2
Coombs	100	96.1	93.1	90.8	86.1	80.7
Black	100	100	100	100	100	100
% of cases where the social utility maximizer coincides with the Condorcet winner	100	92.0	83.5	76.2	64.7	51.1

Table 5.1. The percentage of the cases in which the winners of various procedures coincide with the Condorcet winner (from Merrill, 1984).

Several observations can be made about this table. First, the non-ranked systems - i.e. systems in which no preference orderings over the alternatives are used as the inputs for the determination of social choice - seem to do in general worse than the ranked systems. The only exception to this rule is the performance of the plurality runoff system vis-a-vis the Borda count when the number of candidates is 3. However, for larger numbers of candidates the Borda count gets higher agreement percentages than the runoff method. Secondly, with the increasing number of candidates the agreement percentages of all systems - except Black's of course - become smaller. However, the rate of decrease differs somewhat: it is very fast for plurality voting, whereas for the Borda count the percentage remains at a high level throughout the comparison. Thirdly, in non-ranked systems the runoff method seems to agree more often with the Condorcet winning criterion when k is small, whereas approval voting is the best of non-ranked systems in term of this criterion for larger values of k. Plurality voting is the worst of the non-ranked systems in all but the very lowest values of k. Fourthly, Coombs' and Hare's systems relate to the Borda count pretty much in the same way as the runoff method relates to approval voting, i.e. they do better for small candidate sets while their performance deteriorates vis-a-vis the Borda count when k increases. In sum, while all the systems in Table 5.1. are incompatible with the Condorcet winning criterion, there are considerable differences between the systems as to how common under the random society hypothesis is their disagreement with the Condorcet winner.

Merrill's study sheds some new light on the plausibility of the Condorcet winning criterion as well. It turns out that despite its intuitive plausibility the Condorcet winning criterion does not coincide with another almost equally plausible criterion, viz. social utility efficiency (Weber, 1977). This concept is defined by means of the sum of the individual utility values of the candidates on the assumption that the values are randomly allocated by each voter. The higher the sum for the alternatives chosen under a given system, the higher the social utility efficiency of the system. Now the lowest row in Table 5.1. gives Merrill's data on the percentage of the cases in which the Condorcet winners are also the social utility maximizers for various values of k. Obviously, the two criteria are not identical even though they are far from incompatible.

Unfortunately there are no similar data available on the probabilities of various methods violating the Condorcet loser criterion. Nor do we have probability estimates about the occurrences of incompatibilities with the Condorcet or strong Condorcet conditions for those systems that do not satisfy them. Intuitively, however, the situations in which a system fails on the Condorcet conditions and yet satisfies the Condorcet winning criterion, cannot be very common. On the other hand, it could be argued that these situations are all the more surprising when they occur just because of the intuitive proximity of the Condorcet winning criterion and the two conditions. Be that as it may, the probability estimates must in this case be read with their nature in mind: one can only have probability estimates if one assumes some probability distributions of the relevant phenomena. Assumptions, like that of a random society, are quite crucial in determining the results. To the extent that one feels the assumptions are dubious, one must also take special care in interpreting the results.

The Condorcet criteria and conditions pertain to the pairwise or binary properties of social choice functions. The fact that an alternative is the Condorcet winner does not necessarily guarantee that it performs "positionally" best. If one wants to choose alternatives on the basis of how highly they are ranked on the average in individual preference orders, then the satisfaction of the Condorcet winning criterion or - for that matter - the two Condorcet conditions cannot be used as a guideline for choice. The positional and binary criteria are essentially different even though in some cases there is a relatively straight-forward way of expressing the former in terms of the latter as we have seen in the context of the Borda count. Both types of criteria have some intuitive justification and hence their differences explain to a great extent why procedures that have been constructed with a view to one class of criteria do often quite poorly in terms of the second type of criteria.

To summarize the comparison in terms of the Condorcet criteria and conditions we can present the following table of procedures considered so far:

Procedure	C-winner	C-loser	C-condition	strong C-condition
Amendment	1	1	0	0
Approval	0	0	0	0
Black	1	1	0	0
Borda	0	1	0	0
Coombs	0	1	0	0
Copeland	1	1	0	0
Dodgson	1	0	0	0
Hare	0	1	0	0
Maximin	1	0	1	1
Nanson	1	1	0	0
Plurality	0	0	0	0
Plurality runoff	0	1	0	0
Schwartz	1	1	0	0

Table 5.2. A comparison of voting procedures in terms of Condorcet criteria and conditions.

5.6. The majority winning criterion

Even though the fact that an alternative has more first ranks than any other alternative does not necessarily mean that it would satisfy our intuitions of the socially best alternative as our discussion on Condorcet criteria shows, it still would seem that if a suitably large number of voters rank a given alternative first, our second thoughts concerning the plausibility of choosing the alternative with most first preference ranks will vanish. The reason for this is obvious: when more than 50 % of the voters rank a given alternative first, there is no longer any discrepancy between the Condorcet winning criterion and the plurality winning criterion. Let us define the majority winning criterion as follows: an alternative is the majority winner iff it is ranked first by more than 50 % of the voters. The majority winning criterion is satisfied by a procedure iff it necessarily chooses an alternative that is a majority winner.

The majority winning alternative necessarily coincides with the Condorcet winner when both exist and, moreover, when the former exists the latter also necessarily exists. Hence all procedures satisfying the Condorcet winning criterion satisfy the majority winning criterion. By definition also the plurality, the plurality runoff as well as Hare's and Coombs' procedures satisfy the majority winning criterion. An obvious fact is Black's procedure's compatibility with this criterion as the Condorcet winning criterion is employed first and the Borda criterion only in the absence of a Condorcet winner. If there is a majority winning alternative, it is also the Condorcet winner whereupon it follows that Black's procedure satisfies the majority winning criterion.

The Borda count, on the other hand, does not satisfy this criterion as shown by the following example.

Example 5.12. $|N| = 5$, $X = \{a,b,c\}$

```
3 voters    2 voters
   a           b
   b           c
   c           a
```

a is the majority winning alternative and yet b's Borda-score is the largest.

These kinds of situations could be taken yet further e.g. in the following fashion.

SECOND PROBLEM

Example 5.13. $|N| = 10$, $X = \{a,b,c,d,e\}$

7 voters	1 voter	1 voter	1 voter
a	c	d	e
b	b	b	b
c	d	e	c
d	e	c	d
e	a	a	a

Even though a has a handsome majority of 7 out of 10, it will not be chosen by the Borda count. Moreover, the Borda winner b does not have a single first rank among the 10 voters. The explanation of the discrepancy between the Borda count and the majority winning criterion is, of course, due to the fact that the Borda count utilizes all individual ranks of alternatives, whereas the majority winning criterion counts the first ranks only.

The approval voting also fails on the majority winning criterion.

Example 5.14. $|N| = 5$, $X = \{a,b,c\}$

3 voters	1 voter	1 voter
a	b	c
b	c	b
c	a	a

If the three left-most voters approve both a and b, while the rest approve only their first-ranked alternatives, b wins even though a has the majority of first ranks. Hence the majority winning criterion is violated.

This example can also be taken to the extreme. Consider the following case.

Example 5.15. $|N| = 100$, $X = \{a,b,c\}$

99 voters	1 voter
a	b
b	a
c	c

If all 99 voters approve both a and b, while the remaining one voter approves b only, the winner is b even though 99 out of 100 voters rank a first. This is clearly a major drawback of approval voting.

In this chapter we have mostly been dealing with a set of criteria based on pairwise comparison of alternatives. The only exception is the majority winning criterion and even it can be viewed as a particular strengthening of the Condorcet winning criterion as was just pointed out. Complete successes are procedures satisfying all the Condorcet criteria and conditions, while complete failures satisfy none of them and, moreover, also fail on the majority winning criterion. A glance at Table 5.2. and the discussion in the preceding makes it clear that procedures doing poorly in terms of these criteria are either positional or eliminative. Indeed, it is to some extent a matter of taste whether one calls e.g. Hare's and Coombs' procedures eliminative or multi-stage or positional. What is crucial in determining the outcome is that a large portion of the preference orders of individual voters is needed to make these procedures work. Thus procedures doing badly in terms of Condorcet criteria and conditions are typically ones which look upon the whole preference orders simultaneously and in a sense give "values" to various positions in the individual orders. An exception to this invariance is obviously Nanson's procedure which is intuitively positional and yet does quite well on the Condorcet criteria and consequently also on the majority winning criterion. That procedures which are directly related to binary comparisons, like Schwartz' procedure and the amendment one, do well in terms of the above criteria is to be expected.

But one could build a case for the positional procedures as well, despite the fact that their performance is typically rather poor in terms of the above criteria. For example, one could argue that an alternative which has not been ranked first by any voter should not be chosen even though it might be the Condorcet winner. Similarly one could insist that what is important in the determination of the socially best alternative is not pairwise contests but the number of voters who approve any given alternative. Also the fact that the positional procedures allow the voters to think of the alternatives vis-a-vis each other simultaneously and not in pairwise fashion could be regarded as a virtue of positional procedures although binary ones can also be implemented in a similar fashion. But perhaps more important than any of these arguments is the observation that the Condorcet criteria and conditions are but one - admittedly important - benchmark in evaluating voting procedures. There are others some of which seem at least as compelling as the Condorcet ones. Moreover, as was pointed out in the preceding chapter, the problem of cyclic majorities seems to tip the balance somewhat in favour of the positional methods

SECOND PROBLEM 65

vis-a-vis the binary ones. Anyway, the case for or against either group of procedures is by no means closed yet.

CHAPTER 6

THIRD PROBLEM: HOW TO AVOID PERVERSE RESPONSE
TO CHANGES IN INDIVIDUAL OPINIONS

The notion of winning expressed in the Condorcet winning criterion is compelling precisely because the very concept of winning is so often exemplified by means of pairwise comparisons. It seems difficult to entertain a concept of winning which would not involve defeating some contestant in a binary confrontation. Similarly the criterion discussed in this chapter is intuitively compelling.

6.1. Monotonicity and related concepts

The idea of a democratic form of government seems to imply that governmental actions are related to individual wishes in a plausible way. This idea is incorporated in the notion of responsive government: decisions taken with regard to the people should be responsive to the wishes of the people. If the latter undergo a change, so presumably should the former. Of course, it could be argued that the wishes of the people are not something to be recovered from opinion polls or similar wish-revealing devices, but are something deeper which can at best be inferred from the expressed wishes of the people. Be this as it may, the social choice theory literature shows that even under a minimal construal of responsiveness we may encounter difficulties if the current procedures are resorted to. In other words, even if individuals were fully informed about everything involved in the decisions to be made, there still remains the problem of guaranteeing that their wishes are aggregated into collective choices in a way that in a very modest sense is responsive. The notions related to responsiveness are indeed modest in aiming not much further than at the exclusion of downright perverse reactions of social choices to individual preference changes.

Before going into the precise definitions of concepts related to responsiveness, two intuitive distinctions should be mentioned. Firstly, we shall be dealing with the responsiveness of social choices vis-a-vis the changes of individual wishes or opinions concerning the alternatives at hand. In other words, we shall first look at a preference profile R of voters and notice what the social choice, F(X,R), given these preferences, would be. Then we assume for the sake of argument that certain opinions among the voters change while others are fixed and again we determine the social choices F(X,R'), given these modified preferences R'. The concepts that relate to responsiveness deal with certain types of opinion changes and how they are reflected in social choices.

Secondly, the concepts related to responsiveness can be given both static and dynamic interpretations. Under the former, we shall think of the original and modified preference profiles as belonging to two different voting bodies, while under the latter we shall look at a given voting body and consider a hypothetical preference change in that body. The precise notions to be employed do not distinguish between these two types of interpretation.

Two concepts related to responsiveness stand out in the literature:

(i) monotonicity (or non-negative responsiveness), and

(ii) strong monotonicity (or positive responsiveness) (see Fishburn, 1977; Pattanaik, 1971, pp. 50-52).

The former is intuitively much more plausible than the latter, although both can certainly be considered as desirable properties. Monotonicity requires that, if an alternative x which wins under a given procedure gets more support and nothing else changes in the individual preferences, then that alternative remains the winner after the change as well. It is worth emphasizing that x is not required to be the only alternative in F(X,R) or F(X,R'), that is, x may be one of several tied alternatives. If the only thing that counts in the social choice is to be among the winners, then the requirement of monotonicity can be expressed as follows: it should never hurt an alternative to get more support. Surely this sounds a reasonable requirement. Indeed it seems to lie at the heart of the idea of letting the individuals decide. More formally let a given procedure F be applied to a (X,R)-pair to yield F(X,R) = A as the social choice set (of course $A \subseteq X$). Suppose now that R' differs from R so that the alternative x which belongs to A is ranked higher in one or several individual preferences than in R and that no other changes in preferences have occurred. If for all X, R and R' satisfying these conditions, it is the case that $x \in F(X,R')$, then the procedure realizing F is monotonic.

The condition of strong monotonicity is less compelling although it has some plausibility as well. In May's (1952) original formulation the condition of strong monotonicity is defined for an alternative set which consists of two alternatives only, i.e. for X = {x,y} (Plott, 1976, pp. 558-559). It says that if R is obtained from R' by moving x up in at least one person's preference order and if x was among the winners when the procedure F was applied to R', then x must be the sole winner when F is applied to R. This condition is stronger in the case of two alternatives than the condition of monotonicity because it now excludes the possibility of there being more than one element in F(X,R) even though there may be two elements in F(X,R'). This means that in order to break a tie one needs one preference change only and moreover the winner is the alternative which is moved up in the change in question.

Stated in more precise terms the condition can be defined as follows. The individual i's preferences over X can now take one of the following forms: either $x\ P_i\ y$, or $y\ P_i\ x$ or $x\ I_i\ y$ where I_i is the binary relation of indifference. Denote the first preference by 1, the second by -1 and the third by 0. Let B_i denote the individual preferences so that B_i can take one of the values, 1, -1 or 0, corresponding to $x\ P_i\ y$, $y\ P_i\ x$ or $x\ I_i\ y$. To define the strong monotonicity consider now two n-tuples of individual preferences over X: R and R', and suppose that they are related to each other as follows: $B_i \geq B'_i$ for all i and for some j: $B_j > B'_j$. Suppose now that in this situation F(X,R') = {x} or {x,y}. Then F(X,R) must be equal to {x}.

There are several definitions of strong monotonicity that boil down to May's definition when k=2 (see e.g. Campbell, 1982; Kelly, 1978, p. 44). Many of them operate on social welfare functions instead of social choice functions and are, therefore, not directly pertinent. Pattanaik's (1978, p. 41) definition is, however, more useful for our purposes. Suppose that F is monotonic and suppose that R' is obtained from R - both n-tuples of individual preference orders - so that x which is an element of F(X,R) is moved up in at least one person's preference order vis-a-vis every other element of F(X,R), while its position is no worse with respect to all other elements, i.e. elements in X - F(X,R). Suppose furthermore that for all z and w in X - F(X,R): $z\ R'_i\ w$ iff and only if $z\ R_i\ w$, that is, no changes in individual preferences with respect to other alternatives take place. Then the requirement of strong monotonicity states that x must be the sole element of F(X,R') or the latter is undefined. This rather complicated definition is strictly in the spirit of May's definition. It roughly states that if the requirement of monotonicity is satisfied, then to make a procedure strongly monotonic one needs to impose the requirement that only one favourable preference change is needed to break a tie in favour of that one of the tied alternatives with respect to which the preference change is favourable (i.e. which is moved up vis-a-vis the other alternatives in the tie).

It is obvious that the condition of strong monotonicity is something of a luxury when compared with that of monotonicity. The reason is that the latter has to do with the nonperverse response to individual preference changes in the sense that additional support will not make losers out of winners, while the former deals with the breaking of social ties, i.e. with how much additional support is needed to turn social indifference into a definite position in favour of one alternative. One could well argue that strong monotonicity need not be satisfied, while the failure in monotonicity would seriously undermine the very rationale of trying to aggregate individual preferences. Let us now turn to the evaluation of the voting procedures in terms of these two criteria.

6.2. Successes

The criteria of monotonicity and those named after Condorcet are to a large extent compatible. This applies especially to the Condorcet winning criterion and monotonicity. Intuitively it seems prima facie natural that a procedure which necessarily chooses a Condorcet winner when one exists also satisfies monotonicity. In cases where there is a Condorcet winner, it should not harm it if it is ranked higher than previously. Indeed, if a Condorcet winning alternative improves its position vis-a-vis the other alternatives ceteris paribus, then it certainly remains the Condorcet winner and will thus be chosen by the procedures which necessarily choose the Condorcet winner. However, the procedures satisfying the Condorcet winning criterion typically also specify a choice when no Condorcet winner exists. It is in these cases that monotonicity failures may ensue even though the procedure satisfies the Condorcet winning criterion.

Of the systems considered so far we have observed that the amendment procedure, Copeland's, Dodgson's, Black's, Nanson's and Schwartz' procedures along with the maximin method, satisfy the Condorcet winning criterion. Of these procedures only two, viz. Dodgson's and Nanson's, are not monotonic (see Nurmi, 1983a). And even one of these, viz. Dodgson's method, can be given a construal under which it is monotonic. To wit, if one counts the preference changes in the "binary" fashion as suggested above (4.4.2.) instead of taking into account the secondary preference changes needed to make the alternatives the Condorcet winners, then the method obviously satisfies monotonicity. But as was pointed out above, the construal adopted e.g. by Fishburn (1977) is perhaps closer to the spirit of Dodgson's original proposal. And this construal does not satisfy monotonicity. The following example from Fishburn (1977) shows this.

Example 6.1. $|N| = 19$, $X = \{a,b,c,d,e\}$

5 voters	8 voters	4 voters	2 voters
a	d	b	b
b	a	e	a
c	c	d	d
d	e	a	e
e	b	c	c

To make a the Condorcet winner one would need 3 preference changes from d P_i a to a P_i d with no secondary changes. To make any other alternative the Condorcet winner one would need strictly more changes. Hence a is the Dodgson winner. Suppose now that the group of 2 voters above move a up in their preferences so that their preference orders are:

2 voters
a
b
d
e
c

Supposing that no other voter changes his preference, we observe that now the Dodgson winner is no longer a, but d as only two preference changes are needed to make d the Condorcet winner in this new profile, while a needs three such changes. Thus monotonicity is violated. However, for $|X| = 3$ or 4 Dodgson's method is monotonic (see Fishburn, 1982). We have considered above a 5-alternative case. Hence when the cardinality of the alternative set exceeds 4, the method fails on the monotonicity criterion.

As was pointed out above, Nanson's method is also non-monotonic. An example from Smith (1973, pp. 1036-1037) shows this.

THIRD PROBLEM

Example 6.2. $|N| = 37$, $X = \{a,b,c\}$

10 voters	8 voters	8 voters	3 voters	4 voters	4 voters
a	c	b	c	a	b
b	a	c	b	c	a
c	b	a	a	b	c

The Borda scores are

    a: $14 \times 2 + 12 \times 1 = 40$
    b: $12 \times 2 + 13 \times 1 = 37$
    c: $11 \times 2 + 12 \times 1 = 34$

Thus c is eliminated, and the winner is a. Suppose now that the preference profile is otherwise the same as above but the 3 voters with the preference order cba change their minds in a's favour to cab, and that 3 of the 4 voters with preference order bac change it also in a's favour to abc. The revised preference profile then becomes:

13 voters	11 voters	8 voters	4 voters	1 voter
a	c	b	a	b
b	a	c	c	a
c	b	a	b	c

The new Borda scores are now:

    a: $17 \times 2 + 12 \times 1 = 46$
    b: $9 \times 2 + 13 \times 1 = 31$
    c: $11 \times 2 + 12 \times 1 = 34$

Now b is eliminated and the winner becomes c. Thus Nanson's procedure is non-monotonic (see also Fishburn, 1977; Straffin, 1980, p. 32).

Among the procedures satisfying the Condorcet winning criterion considered here Dodgson's and Nanson's methods are exceptional with respect to the monotonicity criterion. Others are monotonic. In the case of the amendment procedure one must assume that the voting agenda is kept fixed when the choice function is applied to the two preference profiles. If, on the other hand, the agenda depends on, say, the Borda scores of the alternatives, then the monotonicity does not necessarily hold.

In particular, Fishburn (1982) shows that if an alternative's position in the agenda is determined on the basis of a binary relation D of positional dominance so that a precedes b in the agenda if b D a, i.e. b positionally dominates a, then the amendment procedure (or sequential majority elimination procedure as Fishburn calls it) is not monotonic. By definition, b dominates positionally a iff for any rank higher than the lowest, the number of voters who rank b higher than or equal to that particular rank is larger than the corresponding number for a. More precisely, let $p(a_i)$ ($p(b_i)$, respectively) be the number of voters who give a (b) the rank i. The concept of positional domination is defined as follows: b D a iff $\sum_{i=1}^{t} p(a_i) < \sum_{i=1}^{t} p(b_i)$, for the rank values of t ranging from 1 to k-1, where k is the number of alternatives. Now if this dominance relation determines the agenda in the fashion that positionally weaker alternatives are placed earlier in the agenda, then monotonicity is violated.

But in the amendment procedure when the agenda is kept fixed, the monotonicity holds as can easily be inferred. If an alternative is the Condorcet winner, then the increase in its support, ceteris paribus, will certainly keep it that way. If on the other hand, it is not the Condorcet winner, but wins in the amendment procedure on the grounds of being the winner in all last pairwise comparisons, then an increase in its support will obviously not diminish its chances of defeating the alternatives that it defeated with lesser support.

Similarly, it can readily be observed that Copeland's procedure is monotonic: if a Copeland winner is moved up in some persons' preferences and no other changes are made, then that alternative certainly wins all the alternatives (and possibly some new ones) that it won previously. Moreover, it suffers no greater (but possibly rather less) losses than previously. On the other hand, the difference between victories and losses does not change for other alternatives vis-a-vis each other. They may suffer some new losses, though. Thus, they cannot get higher Copeland scores than the previous Copeland winner. Therefore, the latter remains the winner thus proving the monotonicity of the procedure.

Schwartz' procedure is also monotonic. In case there is a Condorcet winner, the procedure chooses that alternative, and if the only admissible change in preferences moves that alternative up then it certainly remains the Condorcet winner and will be chosen. If on the other hand, there is no Condorcet winner to start with, and one of the alternatives in the Schwartz' choice set - say x - is moved higher in some individuals' preferences, ceteris paribus, then it cannot be the case that after the change some alternative which formerly did not beat x would now do so. Consequently, x still belongs to the Schwartz choice set, whereby monotonicity is satisfied.

The maximin winner is the alternative with the highest minimum entry in its row in the pairwise comparison matrix. Of course it need not be unique, but there may be several maximin winners. Suppose now that one of them - say x_i - is moved higher in one or more individuals' preferences. Then there is a change in two entries of the matrix, viz. entries (i,j) and (j,i) will be different from the corresponding entries in the original matrix. The entry (i,j) - i.e. i'th row j'th column - will now be higher than before whereas the entry (j,i) will be lower than before. Hence, the minimum of the i'th row will not be smaller than previously, whereas the minimum of one other row - viz. j'th - may be smaller. All other minima remaining the same, x_i must still be among the winners.

Black's procedure also satisfies the monotonicity condition. In case there is originally a Condorcet winner, it remains the Black and Condorcet winner after it has been moved up in a few preference orders. In case there isn't a Condorcet winner to start with, Black's procedure chooses the Borda winner. If the Borda winner is moved up some preference orders, its Borda score obviously becomes larger than before. Ceteris paribus, i.e. as no other alternative has a higher position than previously and consequently has not a higher Borda score, the original Borda winner remains the Borda winner also after the change in preferences. It may, of course, happen that when in the original preference profile there is no Condorcet winner, there appears to be one when the original Borda winner is moved up. But then this new Condorcet winner must be the alternative which was moved up as no other wins any more alternatives than it did before. This, in turn, means that the alternative which is moved up will be chosen because of its being the Condorcet winner. All the same, it will be chosen, showing that Black's procedure also is monotonic.

Thus, of the procedures we have discussed in the preceding the ones that satisfy the Condorcet winning criterion also satisfy the monotonicity condition with two exceptions: Dodgson's and Nanson's methods are not monotonic. Thus there are several procedures that are complete successes in terms of these two criteria, which intuitively are both very important in the assessment of democratic institutions.

But there are also procedures that fail on the Condorcet winning criterion but satisfy monotonicity. Obviously all our positional methods are monotonic, i.e. the plurality, Borda count and approval voting. (Hare's and Coombs' methods are not classified as positional but as eliminative or multistage, as was pointed out above.) The reason is obvious; as the criterion of winning is positional and the change involved in the monotonicity criterion is a positional improvement, no alternative can be harmed by such an improvement. In particular, if an alternative is the winner of one of these methods, then it remains the winner after the positional improvement.

So, in the case of plurality voting if x wins under a given preference profile, then x will also win under a new preference profile which is obtained from the previous one by moving x up in some persons' preferences, leaving all other things equal. The same is true of approval voting supposing that the changes is preferences do not involve changes in the number of alternatives that the voters deem acceptable, i.e. give their approval to. In the case of the Borda count this is also true as we already suggested above.

So all positional voting procedures fare well in terms of monotonicity because of the nature of the change one focuses upon in determining the monotonicity. In view of the fact that positional and pairwise criteria are different, it is actually quite remarkable how many binary methods also satisfy the monotonicity criterion. On the other hand, we already noticed that none of our three positional methods satisfies the Condorcet winning criterion. Moreover, there is - as we observed - one binary comparison method which is not monotonic, viz. Dodgson's, and one of positional nature - Nanson's - that does not satisfy monotonicity even though it is compatible with the Condorcet winning criterion. Dodgson's and Nanson's methods are, however, by no means the only non-monotonic ones.

6.3. Failures

Perhaps the best-known of the non-monotonic methods is the plurality runoff. An example will show how additional support may turn out to be fatal for an alternative, when this method is used (Straffin, 1980, p. 24).

Example 6.3. $|N| = 17$, $X = \{a,b,c\}$

6 voters	5 voters	4 voters	2 voters
a	c	b	b
b	a	c	a
c	b	a	c

Clearly a wins the runoff between a and b. Suppose now that the two persons whose preference order is indicated as right-most

THIRD PROBLEM

change their minds in favour of a so that the new preference profile is:

6 voters	5 voters	4 voters	2 voters
a	c	b	a
b	a	c	b
c	b	a	c

Now the runoff takes place between a and c. Obviously c wins this contest. Thus the additional support given by the two persons mentioned is fatal for a: it would be chosen without their support but will not be chosen with their support, ceteris paribus. It is easy to see how this oddity emerges: the contestant of a in the runoff is determined by the two voter group. If it votes for a, a's contestant will be c. And since a is not the Condorcet winner in either preference profile, it does not beat all possible contestants it might be faced with in the runoff. In particular, it wouldn't beat c.

Obviously, non-monotonicity has strategic repercussions. If in the previous example (6.3.) the latter preference profile were the "sincere" one, i.e. represented the true opinions of the voters concerning the alternatives, then obviously the campaign-manager of candidate a would be likely to urge the two voter group (which has exactly the same preference order as the 6 voter group) to vote as if their preference order were bac. Thereby a preference profile of the former kind would ensue if no other preference changes were made. In general, when the plurality runoff method is used a group supporting a candidate with almost but not quite 50 % of the first ranks is well-advised to think how to distribute its "surplus" votes in the first round, i.e. the votes that exceed the number needed to get their favourite into the second round.

Examples of a similar kind are not difficult to come by. Indeed, Fishburn (1982) shows that for any two integers P and Q satisfying

(i) $P \geq 2$

(ii) $Q > P+3$ and

(iii) $\frac{(P+Q+4)}{(3P+3Q+3)} < \frac{1}{2}$

we can create a counter-example to the monotonicity of the plurality runoff by assuming that the voters are distributed over the preference orders first as follows:

Example 6.4.

Q voters	Q voters	Q voters	P+2 voters	P voters	P+1 voters
a	c	b	b	c	a
b	a	c	a	b	c
c	b	a	c	a	b

and then as follows:

Q+3 voters	Q+2 voters	Q voters	P-1 voters	P-2 voters	P+1 voters
a	c	b	b	c	a
b	a	c	a	b	c
c	b	a	c	a	b

In the first case the runoff takes place between a and b, as obviously Q+(P+1) > Q+P and Q+(P+2) > Q+P. In this runoff a wins because of conditions (i) and (ii).

Now the latter preference profile is obtained from the first by moving 3 persons from the preference order bac to the order abc and 2 persons from the order cba to cab. Both of the changes involve nothing more than moving the winner a up in some preference orders. Yet the winner does not remain the same after the change as the runoff now takes place between a and c, and in this contest the winner is c. This can be seen from the following. Firstly, the distribution of first preferences is:

$$a: \quad Q+3+P+1 = Q+P+4$$
$$b: \quad Q+P-1$$
$$c: \quad Q+2+P-2 = Q+P$$

Obviously, the runoff takes place between a and c. The difference between their first rank numbers is 4 in a's favour. Now in the runoff a gets P-1 votes from the voters who have b as their first preference, while c gets Q votes. The difference between these new votes is now larger than 4 in c's favour: Q - (P-1) is by assumption strictly larger than (P+3) - (P-1) which equals 4. Hence c wins the runoff contest.

A more general result pertaining to the monotonicity of the elimination procedures is also given by Fishburn (1982). It states that for X = {a,b,c} whenever it is the case that a procedure satisfies the following condition:

{a D b and c D b and n(a,c) > n(c,a)} => F(X,R) = {a}

THIRD PROBLEM 77

the procedure is not monotonic. In other words, if the procedure behaves so as to choose among the two alternatives that both positionally dominate the third the one which is preferred to the other by more persons than vice versa, then the procedure violates monotonicity. Obviously the plurality runoff behaves in this fashion and thus falls within the realm of the result.

One can also see that Hare's procedure behaves in the same way as the plurality runoff in Example 6.4. That is, in the first case c is eliminated as it has the smallest number of first ranks. In the contest between a and b, a wins. In the latter profile, the winner is c as in the plurality runoff. Thus, Hare's procedure is also non-monotonic. The reader can easily verify that Coombs' procedure would also make the same choices as the plurality runoff and Hare's procedure in Example 6.4. Hence it fails on monotonicity as well.

What about the strong monotonicity or positive responsiveness of the procedures? Now obviously only those procedures that are monotonic can also be strongly monotonic. That the simple majority rule defined for the two alternative case only is strongly monotonic is known from May's (1952) axiomatization. To the extent that the concept of a tie can be given the interpretation that the tied alternatives - necessarily two in number - receive an equal number of votes in the last pairwise comparison, the amendment procedure is also strongly monotonic. This follows from the fact that if one of the voters changes his mind about the priority of the alternatives, then one of the tied alternatives gets one vote more than previously while the other one loses one vote. The difference in the total number of votes is thus two.

Copeland's procedure, on the other hand, is not strongly monotonic. If two alternatives, x and y, have the same difference in the number of victories and losses, then the fact that some person who formerly preferred x to y changes his opinion to y P_i x, does not necessarily make y the sole winner. The reason is that the preference change in itself need not affect the number of victories and losses of x and y. The alternative y may well be the majority winner in both cases.

By a similar process of reasoning it can be seen that Schwartz' procedure is not strongly monotonic. For instance, if there is a top cycle through x, y and z, then the fact that x is moved up in some persons' preferences vis-a-vis y and z does not necessarily break the cycle. The reason for the fact that strong monotonicity is violated is that the sizes of the majorities in each link of the cycle play no role in the determination of whether there is a cycle or not. On the other hand, the preference changes affect the sizes of the majorities and thus do not necessarily reverse the direction of the majority dominance. That the maximin method also fails on strong monotonicity is obvious as the only relevant thing in that method is the smallest amount of support given to each of the alternatives. A change in the preference concerning a pair of tied alternatives does not necessarily affect the smallest amount of support given to any of the tied alternatives.

The plurality procedure pays attention to the first preferences of the voters only. Hence, if there is a tie between several alternatives, a change of preference between some of them does not break the tie unless the change concerns the alternatives reported as first preferences. Thus, the plurality method also fails on strong monotonicity. So does approval voting. Only if the preference change involves a change in approvals can approval voting ties be broken by single voters.

The Borda count, in contrast, seems to be strongly monotonic, as obviously moving up one alternative of a pair of tied ones, ceteris paribus, increases the Borda score of that alternative by one point and possibly decreases the score of its contestant by one point.

In Black's procedure a tie can only occur when there is no Condorcet winner. In such a case the Borda winners are chosen and a tie may occur. If after the preference change of one individual in favour of one of the tied alternatives there still is no Condorcet winner, then obviously the procedure is strongly monotonic as is the Borda count. On the other hand, if after the preference change there is a Condorcet winner, then it must be the particular alternative that was moved up in an individual's preference order, ceteris paribus. Hence, in this case also the condition of strong monotonicity is satisfied.

6.4. The relevance of the monotonicity criteria

The monotonicity criterion is undoubtedly one of the basic criteria of democratic group decision making. The idea of counting votes in an effort to determine group preferences assumes that the more support an alternative has, the better chances it has to be chosen as the socially most preferred alternative. As we have seen in procedures lacking the monotonicity property, additional support may turn out to be harmful for an alternative. Thus, one can build a strong case for monotonic procedures in contrast to non-monotonic ones.

On the other hand, the condition of strong monotonicity does not seem equally important. Indeed, one could even argue that it is implausible because to break social ties, one person's change of opinion does not seem intuitively adequate. Of course, one could counter this objection by asking how many preference changes then would be adequate to turn the social indifference between two or more alternatives into a victory for one of them. Be that as it may, the condition of strong monotonicity clearly seems less compelling than that of monotonicity simpliciter.

In one special situation one could make a case for strong monotonicity using Pareto optimality (to be discussed in the next chapter) as the background. When the entire society is indifferent as to whether x or y should be chosen and then one person changes his mind so that his preference becomes $x\ P_i\ y$, then the Pareto condition would require that y be not chosen. Strong monotonicity would in this case require that x be chosen. Surely these two requirements are not incompatible. But in general the conditions deal with entirely different types of situations: when there is a social tie between two alternatives, the fact that one person is needed to break such a tie by no means implies that either of the alternatives were excludable in terms of the Pareto criterion.

Comparing the above considerations with those made in the preceding chapter we observe that the Condorcet criteria and the monotonicity criteria are not mutually exclusive. The amendment procedure, Copeland's, Schwartz' and Black's procedures are all monotonic and satisfy the Condorcet losing and winning criterion. If we take into account the Condorcet conditions and strong monotonicity, the picture changes somewhat as the only procedures satisfying strong monotonicity are the amendment procedure along with the Borda count and Black's method.

It could be argued that monotonicity violations, although always serious when they occur, are probably not very common in practice. Unfortunately there are no studies reporting the probability estimates of violations under various assumptions. What is certain, however, is that monotonicity violations are often deliberately exploited in political manoeuverings. For instance, when the plurality runoff system is utilized the supporters of a very strong candidate often decide to give their "excess" votes on the first round not to their favourite but to the one of his possible contestants whose presence in the second round would best guarantee the victory of their favourite. For instance in a situation like in Example 6.3. suppose that the latter preference profile is the true one. This means among other things that 8 persons have the preference order abc. However, this group is well advised not to give all their votes to a in the first round as this would lead - assuming that the other voters vote according to their true preferences

- to the second round contest between a and c, and this again would result in the choice of c, the worst alternative in the group's opinion. Therefore, by giving 2 votes to b in the first round this group would guarantee a runoff contest between a and b whereupon a would win. There is anecdotal evidence that this kind of manoeuvering is well-known to people involved in practical politics. For instance, before the previous (1982) presidential elections in Finland, there were rumours of this kind of strategic or sophisticated voting being planned in the electoral college. Even though non-monotonicity is by no means a necessary - let alone a necessary and sufficient - condition for successful strategic voting, it is certainly an invitation to it.

CHAPTER 7

FOURTH PROBLEM: HOW TO HONOUR UNANIMOUS PREFERENCES

Monotonicity thus requires that it should never be harmful for an alternative if its support increases. Now if the support increases so as to reach unanimity, it would seem reasonable that the group should have its way precisely as it wishes. But what does unanimity mean in group decision making? Of course, if the group consists of n persons with exactly identical preference orders, then unanimity would be a straight-forward notion. But supposing that the group is not entirely homogeneous, but there are some alternatives with respect to which unanimity prevails, how could such an agreement of opinions be taken into account in social choices in a plausible way?

7.1. Unanimity and Pareto conditions

Suppose that there are two alternatives x and y in X for which every voter has an identical preference x P_i y. Then it would seem natural to exclude the choice of y. Or as Plott (1976, p. 528) puts it: "if x is available and everyone ranks x above y, then y should not be chosen".

It should be observed that the consequence does not state that x should be chosen. Moreover, the antecedent is a rather stringent requirement that everyone has an identical strict preference between x and y. This condition is called the Pareto principle by Plott. More specifically it is the weak Pareto condition (Kelly, 1978, pp. 10-11). If one wants to emphasize the nature of the underlying unanimity, it can also be called unanimity simpliciter (Fishburn, 1973, p. 83).

A slightly more general concept is that of the strong Pareto condition. This condition states that if every i in N regards x at least as good as y and at least one member of N deems x strictly better than y, then it should not be the case that y is chosen. In other words, if $(\forall i)(xR_i y)$ and $(\exists i)(xP_i y)$, then $y \notin F(X,R)$. In this case unanimity concerns nonstrict preferences, i.e. everyone deems x at least as good as y. But in addition to this there must be at least one person whose strict preference is xP_iy. This condition is also called strong unanimity (Fishburn, 1973, p. 83). These two principles then both exclude the choice of y under somewhat different conditions. One could easily introduce more unanimity-related principles of choice. The following two seem particularly pertinent and intuitively plausible.

Firstly, suppose that the antecedent conditions of the weak Pareto condition are satisfied. One could then require that x would also collectively be regarded at least as good as y. Let us call this condition the conservative weak Pareto condition. It states then that if everyone regards x as strictly better than y, then the social choice should be either x or x and y or neither should be among the chosen alternatives. This differs from the weak Pareto condition in allowing for y to be chosen but only when x is chosen as well.

Secondly, the strong Pareto condition could also be modified or generalized in a similar fashion, that is, its consequence could also be that x or x and y or neither should be chosen. This generalized condition could in analogous fashion be called the conservative strong Pareto condition. It states that if everyone regards x at least as good as y while at least one person regards x as strictly better than y, then x or x and y or neither should be in the social choice set. More formally, the first of the two conditions says that if for all i in N, $x P_i y$, then it is not the case that $y \in F(X,R)$ and $x \notin F(X,R)$ (conservative weak Pareto). If, on the other hand, for all i in N, $x R_i y$ and for at least one j in N, $x P_j y$, then it is not the case that $y \in F(X,R)$ while $x \notin F(X,R)$ (conservative strong Pareto).

The Pareto conditions are prima facie plausible in the sense that they seem to exclude downright perverse social responses to individual preferences. Obviously the conservative Pareto conditions are much milder requirements than the ordinary weak and strong Pareto conditions. What is perhaps worth emphasizing is that when $|X| = 2$, the strong monotonicity and strong Pareto condition are related to each other. If $F(X,R)$ is not constant, then strong monotonicity implies the strong Pareto condition (Fishburn, 1973, p. 25). So these two classes of conditions are not incompatible, although to what extent they coincide in the voting procedures we have been discussing remains to be seen.

Another feature worth observing is that there is no obvious relationship between the Pareto conditions and the Condorcet criteria. The reason for this is rather straight-forward: Pareto conditions are defined with respect to a decision rule which is

FOURTH PROBLEM   83

unanimity, whereas the Condorcet criteria deal with the simple majority rule only.

7.2. Successes

Now obviously the strong Pareto condition implies the weak one in the sense that if a voting procedure satisfies the former, it necessarily also satisfies the latter. On the other hand the converse is not necessarily true, i.e. if a procedure satisfies the weak Pareto condition, it does not follow that it also satisfies the strong one.

The complete successes in terms of the Pareto conditions are two positional voting procedures, viz. plurality voting and the Borda count. If the former is utilized and x is ranked at least as high as y by all voters and strictly higher by at least one voter, then obviously x gets more votes than y (assuming that the voters vote according to their preferences). Consequently, y is not chosen if x is available. This means that the plurality procedure satisfies the strong Pareto condition. Hence it also satisfies the weak and conservative Pareto conditions.

Similarly the Borda count will also assign to an alternative x which is ranked at least as high as y by all voters and by at least one of them strictly higher, the Borda score which is strictly larger than that of y. Hence y is not chosen if x is feasible. Thus, all the above Pareto conditions are satisfied.

Turning to the binary comparison procedures, we notice that Copeland's procedure satisfies the strong Pareto condition because if x is ranked at least as high as y by all voters and strictly higher by at least one of them, then x obviously beats y in the pairwise comparison and, moreover, defeats all those alternatives that y beats. Furthermore, x cannot possibly lose for a larger number of alternatives than y. Consequently, y is not chosen if x is available. Obviously, then, the other Pareto conditions are satisfied as well.

Consider now Dodgson's method and suppose that the antecedent of the weak Pareto condition is fulfilled for x and y, i.e. x is strictly preferred to y by all voters. This means then that x can be made the Condorcet winner with strictly fewer preference changes than y. Hence, y will not be chosen by Dodgson's method. Suppose now that the antecedent of the strong Pareto condition is satisfied. Then to make y the Condorcet winner one needs all the preference changes as one needs to make x the Condorcet winner plus at least one additional change, viz. for the only voter who has strict preference for x over y. Hence, y cannot be chosen in this situation. Thus, the strong Pareto condition is satisfied by

Dodgson's procedure as well. A fortiori, the conservative Pareto conditions are satisfied.

The maximin method satisfies the strong Pareto condition as obviously the number of voters strictly preferring x to z must be at least as large as the number of voters strictly preferring y to z, that is, $n(x,z) \geq n(y,z)$ for all $z \in X-\{x,y\}$, assuming that the antecedence of the strong Pareto condition is satisfied, i.e. that x is ranked at least as high as y by all voters and strictly higher by at least one of them. Now clearly x is y's toughest competitor because $n(y,x) = 0$, surely the lowest minimum score available. It might be possible that also for x there is an alternative, say z, such that $n(x,z) = 0$, but this would only mean that x is not chosen either. This is because there is always at least one alternative v such that $n(v,w) > 0$ for all other alternatives w. Hence in case the antecedence of the strong Pareto condition holds for x and y, y cannot be chosen because y's score is lower than that of at least one alternative (see also Kramer, 1977 and Fishburn, 1977). From the strong Pareto condition all the other Pareto conditions follow as we have just seen.

Black's procedure also satisfies the strong Pareto condition. If there is a Condorcet winner, then this cannot be y assuming that the antecedent of the strong Pareto condition holds because x defeats y in a pairwise comparison, i.e. $n(x,y) > n(y,x) = 0$. Hence y is not chosen by Black's procedure if there is a Condorcet winner. Nor is y chosen if - in the absence of a Condorcet winner - the Borda winner must be chosen. This follows from the fact that the Borda count satisfies the strong Pareto condition. Hence in either case, the strong Pareto condition is satisfied.

It can easily be verified that the plurality runoff method also satisfies the strong Pareto condition. Suppose that for all i in N: $x R_i y$ and for some j in N: $x P_j y$. Then if x is ranked first by more than 50 % of the voters, it will obviously be chosen and y will not be chosen. If on the other hand x is not ranked first by more than 50 % of the voters, it is possible that it will not be chosen for the second round. If this is the case, then it is clear that y also will be left out of the runoff contest, as it cannot have more first ranks than x. If finally x is chosen for the runoff contest, then y cannot be the winner because x will defeat it should it manage to get to the second round with x. In all cases y is not chosen, whereby the strong Pareto condition is satisfied.

Nanson's method also satisfies the strong Pareto condition. This is due to the fact that at no stage in the elimination process does x get a smaller Borda score than the alternative y if $x R_i y$ for all i in N. Hence, if y survives to the final stage in the process, then it will be eliminated in favour of x in the final elimination. In any case, it will not be chosen.

Let us finally turn to Hare's and Coombs' methods. In neither of them can y be chosen if x R_i y for all i in N because in order to be chosen an alternative must have more than 50 % of the first ranks among the remaining alternatives. Obviously, if x still survives, y cannot have strictly more first ranks. And if y survives, so must x because in Hare's procedure the elimination takes place according to the number of first ranks an alternative has. Hence y is necessarily eliminated at some stage of the process.

In Coombs' procedure, on the other hand, the elimination criterion is the number of last ranks. Obviously, x cannot have more of them than y. Indeed, as long as y is present in the process, x has strictly fewer last ranks than it. Hence both Hare's and Coombs' procedures satisfy the strong Pareto condition.

The strong Pareto condition and, consequently, the weak and conservative Pareto conditions are relatively common among the voting procedures we have been investigating. They are not, however, universal. Indeed, there is a procedure which is widely used and which does not satisfy any Pareto condition mentioned in the preceding.

7.3. A partial failure and a total failure

The performance of approval voting with respect to the Pareto conditions could be called a partial failure. The following example shows that the procedure does not satisfy the weak Pareto condition.

Example 7.1. |N| = 3, X = {a,b,c}

1	2	3
a	a	c
b	b	a
c	c	b

Suppose now that persons 1 and 2 approve both a and b whereas person 3 approves c only. If they all vote accordingly, the winners are a and b. Consequently, it is not the case that b is not chosen even though everyone prefers a to b. Hence the weak Pareto condition is not satisfied.

By a modus tollens argument we can infer that the strong Pareto condition is not satisfied either.

if p is true, then q is true
q is not true
therefore p is not true.

Substituting "the procedure satisfies the strong Pareto condition" for p and " the procedure satisfies the weak Pareto condition" for q, we see that the above inference is, indeed, valid.

However, the conservative Pareto conditions are satisfied by approval voting. That is, if x is ranked at least as high as y in everyone's preference list and strictly higher in at least one persons' preference order, then obviously x gets at least the same number of approvals as y. Hence it cannot be the case that y is chosen while x is not. Consequently, the strong conservative Pareto condition is satisfied. It follows then that the weak conservative Pareto condition is also fulfilled.

Another partial failure with respect to the Pareto conditions is Schwartz' procedure. The following example demonstrates its failure on the weak Pareto condition.

Example 7.2. $|N| = 3$, $X = \{a,b,c,d,e\}$

1	2	3
a	c	b
b	d	c
c	a	d
d	b	a
e	e	e

Now obviously, $F(X,R) = \{a,b,c,d\}$ as every proper subset of this choice set contains at least one alternative which is defeated by some alternative not in the choice set. However, c is preferred to d by all voters and yet d is chosen along with a, b and c.

The conservative Pareto conditions are, however, satisfied by Schwartz' procedure. To wit, if the Pareto dominated y is chosen, then the Pareto dominant x must be chosen as well because otherwise there would be an alternative - viz. x - that defeats an alternative - viz. y - in the choice set. Because the strong conservative Pareto condition is satisfied, so is the weak conservative one.

If the performance of Schwartz' method and approval voting leaves something to be desired, the showing of the amendment procedure is extremely poor in terms of the Pareto criteria. The following example demonstrates this (Nurmi, 1983a, p. 196).

FOURTH PROBLEM

Example 7.3. $|N| = 3$, $X = \{a,b,c,d\}$

1	2	3
a	c	b
b	d	c
c	a	d
d	b	a

Consider now the agenda: 1. b vs. c, 2. the winner of 1. vs. a, and 3. the winner of 2. vs. d. Then the winner of the first ballot is b, the winner of the second ballot a, and the winner of the third one d. Hence the over-all winner is d. However, c is preferred to d by all voters. This shows then that the weak Pareto condition is violated. Consequently, the strong Pareto condition also is not satisfied. The example shows furthermore that the amendment procedure fails on the conservative Pareto criteria because it is not the case that c is chosen along with d.

7.4. Relevance and compatibility with other criteria

From the very beginning Pareto conditions have played a prominent role in social choice theory. For instance in the famous Arrow impossibility theorem (in its 1963 version) Pareto optimality features as one of the conditions which turn out to be incompatible. Similarly, Sen's liberal paradox consists of showing the incompatibility of Pareto optimality and what he calls minimal liberty (Sen, 1970, pp. 87-88; see also Sen, 1983). It would make little sense to show the incompatibility of rather implausible theoretical properties. Thus, the Pareto conditions have apparently been deemed rather plausible properties of social choice procedures. And, indeed, one would certainly be hard pressed to maintain that there is nothing wrong in a procedure that results in the choice of y even though everyone in the voting body strictly prefers x to y. As a matter of fact, one could even go as far as to argue that such a procedure is not democratic in the basic sense of the term because the choice seems to be dictated by something other than the individual preferences which happen to agree on this pair of alternatives. In the amendment procedure which is the only total failure on the Pareto conditions, this something is the agenda, i.e. the order in which the alternatives are brought into pairwise comparison. Riker (1982, pp. 117-118) argues that the intuitive force or plausibility of Pareto optimality stems at least partly from the observation that this criterion carries within itself two others, viz. monotonicity and non-imposition.

More specifically, Riker maintains that monotonicity and non-imposition imply Pareto optimality. Although monotonicity and Pareto optimality are plausible and non-imposition even more so, Riker's view of their relationships is correct only insofar as one is dealing with social welfare functions instead of social choice functions.

Non-imposition or citizens' sovereignty is defined as follows (see Riker, 1982, p. 117): the social choice is imposed if a fixed alternative x is the winner for all preference profiles of voters. Now obviously the fact that the amendment procedure is incompatible with Pareto optimality does not follow from its incompatibility with either non-imposition or monotonicity as we have already observed that the amendment procedure is monotonic. It also satisfies non-imposition because by constructing a preference profile so that an alternative, say x, is the Condorcet winner we can guarantee that it will be chosen by the amendment procedure. As this can obviously be done for any alternative, there is a fortiori no way in which an alternative can become a winner regardless of the preference profile. Consequently, the intuitively undemocratic feature in the failure of some procedures on the weak Pareto criterion is conceptually distinct from their performance with respect to the monotonicity and non-imposition criteria.

The same observation can also be made about the two other procedures that fail on the weak Pareto criterion: Schwartz' procedure and approval voting. Both of them satisfy monotonicity and quite obviously also the non-imposition criterion. Yet they are incompatible with the weak Pareto (although not with the conservative) criterion.

Could one then argue that a failure on either monotonicity or non-imposition is *sufficient* for non-Pareto optimality? The answer is no as far as monotonicity is concerned. Several elimination procedures (e.g. plurality runoff and the Hare system) fail on monotonicity and yet satisfy Pareto optimality. On the other hand, if a procedure is imposed it clearly violates all Pareto criteria: if x is the winner regardless of the preference profile, then it is, in particular, the winner when the preference profile is such that all voters prefer y to x. Hence none of the Pareto criteria is satisfied.

Now assuming then that Pareto optimality is really something different from the previous criteria we have been using, it is natural to ask whether it is in the end a desirable property. As was pointed out above, the Pareto criteria are present in the well-known impossibility results. Thus their removal would certainly make things look a lot better for an institution-designer. Let us briefly look at some of the results in which the Pareto criteria feature.

As was already mentioned, Sen's (1970) result essentially establishes the fact that individual liberty and collective rationality are incompatible insofar as the latter is defined as Pareto optimality and the former in a very minimal sense. It seems plausible to argue that individual liberty means that the individual has a personal sphere in which his preference is the sole determinant of the social preference (Sen, 1983, p. 7). Minimal liberty, on the other hand, requires that there is such a sphere for at least two individuals. In other words, it would seem reasonable to say that whenever a person has some liberty it implies that his own will dictates whatever is being done in affairs within that realm. Sen's concept of minimal liberty requires that there is such a personal sphere for at least two individuals. Surely this requirement would hardly be enough if one were looking for a sufficient condition for a libertarian society, but it is minimal all right and in that sense a necessary condition for any definition of a libertarian society. Sen's result, however, states that even this minimal construal of liberty is incompatible with weak Pareto optimality and an unrestricted domain of social choice. That is, if one looks for a procedure that is general in the sense of resulting always in a non-empty choice set, then one will either have to do without weak Pareto optimality or without minimal liberty.

Elster (1978, pp. 40-41) gives an example of Sen's result which has also been called the liberal paradox. Consider two local communities A and B and a situation which concerns the establishment of a nuclear plant. Let the action alternatives be the following:

(a) the plant will be built in A,
(b) the plant will be built in B, and
(c) there will be no nuclear power plant in either community.

Suppose now that the representatives of A and B have the following preference orders:

A	B
b	c
a	b
c	a

These orders could e.g. follow from the conviction of representatives of A that nuclear power is essential for the community but that radiation may cause problems for the local population. Representatives of B, on the other hand, would be most pleased if there were no nuclear plant, but if one has to be established, they would rather have it in their own community because of the employment opportunities it creates. Suppose moreover that the communities agree that in this choice situation the condition of liberalism should be upheld in the sense that if there is a choice between building a plant in community x and not building a plant at all, community x should determine the social preference between these two alternatives. Let us also assume that the representatives feel that Pareto optimality should be honoured, i.e. if the representatives of both communities prefer a given alternative to the other, the unanimous preference should prevail.

Under these assumptions we have a social preference cycle: a P c (by liberalism because representatives of A decide this preference), b P a (by weak Pareto optimality) and c P b (by liberalism because representatives of B decide this preference). Thus, no matter which outcome is picked, there is always an outcome socially preferred to it.

The point of the example is to show that it is possible to find a preference profile such that weak Pareto optimality and minimal liberalism are incompatible. It could be claimed that the artificial two-community case is a special one and as such not convincing. The example can, however, be easily expanded e.g. into the following profile:

A	B	C	D
b	.	.	.
.	.	.	.
.	.	.	.
.	c	b	c
.	.	.	.
a	.	.	.
.	b	a	b
.	.	.	.
.	.	.	.
c	.	.	.
.	a	c	a
.	.	.	.
.	.	.	.

With any number of alternatives and persons in preference profiles C and D we get the liberal paradox provided that persons A and B are considered to be entitled to their private spheres of choice so that A decides society's preference between a and c while B decides its preference between c and b. Sen's result thus states that the requirements of weak Pareto optimality, minimal liberty and an unrestricted domain of social choice are incompatible, i.e. there is no such procedure that would simultaneously satisfy all these requirements. Surely the emphasis on collective rationality at the expense of individual liberty seems to have gone a bit too far if the former is understood as Pareto optimality, as minimal liberty is indeed a very mild requirement. From this perspective the Pareto criteria do not seem plausible at all.

On closer inspection, however, the liberal paradox per se does not constitute a convincing case against the weak Pareto optimality. In the usual construals of the paradox, like in Elster's version presented above, the individual preferences are meddlesome. Thus, e.g. the representatives of A prefer the construction of the plant in community B to its construction in A. If these types of meddlesome preferences were excluded, one could find schemes that satisfy the weak Pareto optimality and the unrestricted domain (see Sen, 1976; Suzumura 1983, pp. 180-200).

Moreover, one could argue that the liberal paradox confounds two issues that would naturally call for separate handling, viz. the issue of whether there should be a nuclear plant in the first place (i.e. c vs. not c) and - if the choice is not c - where it should be located (i.e. a vs. b). If these two issues were decided separately, then no difficulty with the weak Pareto optimality would seem to ensue. Finally, Kelsey's (1985) theorem states that Sen's impossibility result can be strengthened so that, instead of the weak Pareto optimality, one of its implications, viz. non-imposition, can be shown to be incompatible with minimal liberty and the unrestricted domain. All these considerations, thus, suggest that the liberal paradox in itself is insufficient to cast doubt on the plausibility of the weak Pareto optimality.

CHAPTER 8

FIFTH PROBLEM: HOW TO MAKE
CONSISTENT CHOICES

As was mentioned above the Pareto optimality criterion is sometimes regarded as a group rationality criterion. In other words, the choice behaviour which is consonant e.g. with the weak Pareto criterion is considered rational from the group's point of view in an analogous fashion as it would seem rational for an individual to choose from the set {x,y} the alternative which he strictly prefers. Group or collective rationality is, however, a notion that has been employed in other meanings as well. In these other interpretations it usually refers to a sort of consistency in choice behaviour or, more specifically, to the invariance of choice sets when some transformations are made to the alternative sets or preference profiles. Let us now consider these interpretations of collective rationality in some detail.

8.1. Choice set invariance criteria

There are basically two approaches to the analysis of consistency of the choice behaviour of a group. One of them deals with what is called intra-profile choice set changes and the other with inter-profile ones. The former focuses on what can happen to the social choice sets when the preference profile is fixed due e.g. to the fact that the decision making body and the opinions of its members do not change during the analysis. That changes can under these circumstances occur in social choice sets, may prima facie seem strange. On closer inspection, however, it turns out that the behaviour of various choice procedures is rather interesting and even unexpected when one considers the various subsets of the alternative set, on the one hand, and the entire alternative set, on the other. It is this kind of comparison that the intra-profile analyses are concerned with.

The inter-profile changes in turn involve changes in social choices when the preference profiles change e.g. due to the fact that there are different groups making choices from the same set

FIFTH PROBLEM

of alternatives. The point in developing and using inter-profile choice set invariance criteria is that even when the groups are different it would seem reasonable to expect a certain conformity in their choice behaviour in specific contexts. Similarly, the rationale of the intra-profile criteria is that a certain invariance in social choices would seem intuitively "natural" given the invariance of preference profiles.

The specific question asked is: does the fact that not all alternatives are considered simultaneously affect the social choices realized by the procedure in question? In other words, are the winners of the procedure "genuine" ones in the sense that adding "artificial" alternatives and/or changing the order in which the alternatives are brought into consideration do not affect the winners? The inter-profile approach in turn asks what kind of effects the partitioning of the *voter set* may have on the outcomes. The intra-profile approach is similar in its general outlook to the analysis of procedures in terms of Pareto criteria or Condorcet criteria, whereas the inter-profile approach is somewhat reminiscent of monotonicity-type considerations. The choice set invariance properties are, however, basically different from these other properties.

Of the inter-profile properties the best-known is undoubtedly consistency (see Young, 1974). It requires that if two distinct groups applying a given procedure to the same alternatives come up with at least partly overlapping choice sets, then if the groups are combined and nothing changes in the voters' preferences or alternative set, the groups should choose exactly those alternatives their choice sets had in common, assuming that the same procedure is used all along. As a special case consider one in which two groups using a given procedure both choose an alternative, say x. Then obviously there is an overlap in their choice sets. The consistency requirement now states that if the procedure were applied to a situation in which the two groups consider the alternative set simultaneously, they should also come up with x and with x only.

Slightly more formally, let there be two groups N_1 and N_2 of voters so that no individual belongs to both groups. For a given set X of alternatives, let the preference profiles of the groups be R^1 and R^2, respectively. Consider now a procedure that realizes the social choice function F and suppose that $F(X,R^1) \cap F(X,R^2) \neq \emptyset$. Let the same procedure be applied to the same alternative set so that the groups - or rather the preference profiles - are combined, i.e. we assume that the group $N = N_1 \cup N_2$ makes the choice and no changes take place in individual preferences. The consistency property now requires that the alternatives chosen by

N be exactly those that were chosen by both N_1 and N_2 (assuming, as we did, that at least one particular alternative *is* chosen by both). In other words, we may say that a procedure realizing F is consistent iff whenever $F(X,R^1) \cap F(X,R^2) \neq \emptyset$, it is also the case that $F(X,R) = F(X,R^1) \cap F(X,R^2)$, where R is the preference profile of N.

As it stands the consistency property seems a very natural requirement. What it says is that if something is considered best by two distinct groups of voters, then that same something should also be considered best by the group consisting of these two groups assuming that the individual preferences remain the same throughout. One must, however, pay due attention to fact that consistency actually requires more than what has been just stated, viz. the alternatives chosen by the combined group N must be exactly the same as the alternatives that both groups choose when acting separately. If the groups choose entirely different things so that no alternative belongs to both choice sets, then consistency does not say anything, i.e. no matter what the combined group chooses under this assumption, the behaviour is compatible with consistency.

The fact that the groups choose some common elements implies that the intersection set must not contain more or less alternatives than the set chosen by N. Intuitively one could deem this as unnecessarily stringent; sometimes the requirement $F(X,R^1) \cap F(X,R^2) \cap F(X,R) \neq \emptyset$ would seem satisfactory. Certainly the violations of consistency that do not even satisfy the latter requirement are more serious than the ones that nonetheless are compatible with it. We shall, however, stick to the stricter notion of consistency and call the procedures that satisfy the latter requirement weakly consistent. Obviously, consistency implies weak consistency, i.e. every procedure satisfying the former also satisfies the latter, but not vice versa.

Of intra-profile properties let us first introduce the weak axiom of revealed preference (WARP) discussed by Arrow (1959). Consider a proper subset of the alternative set X, say A. If the large set X consists of e.g. construction project alternatives, the subset A would consist of those projects that cost at most k million pounds. Suppose now that the procedure F is used in the voting body with the result $F(X,R)$. Suppose, moreover, that $F(X,R) \cap A \neq \emptyset$, i.e. that when considering all projects the group decides that some projects in A are among the best. In other words, the group regards some of the projects in A to be in the class of best alternatives. Now of course the group could also consider projects in A only instead of all projects. For instance, the overall budget could restrict the set of realistic alternatives

FIFTH PROBLEM

to the set A only. Suppose that the same procedure, when applied to A, would yield F(A,R), i.e. the best alternatives in A are F(A,R). Now, the procedure satisfies WARP iff F(X,R) ∩ A = F(A,R). This means that we should come up with exactly the same outcomes via two routes:

1. by applying the procedure to X and taking the intersection of the chosen alternatives with A, and

2. by applying the procedure directly to A. Here we assume that route 1. leads to a nonempty intersection.

Plott (1976, p. 550) observes that WARP actually contains two requirements:

(i) that the winners in X are winners in every subset of X to which they belong, and

(ii) if there is a tie between some alternatives in a subset of X, then either all the tied alternatives or none of them will be among the winners in X.

The former requirement sounds quite reasonable. Indeed, the very notion of "winner" seems to imply it. The latter, on the other hand, is less plausible as it is often the case that the performance of one of the tied alternatives in a subset of X, in competition with the others, makes it intuitively "best" when the entire X is considered.

In Figure 8.1. have a situation in which WARP is not violated. The choice set from X' coincides precisely with the intersection of the choice of X and X'. If this holds for any subset of X (and not just for the one we have depicted), the procedure determining the choice sets satisfies WARP. Thus to prove that a procedure satisfies WARP one has to go through all possible subsets of X and show that the situation is in each case similar to that in Figure 8.1.

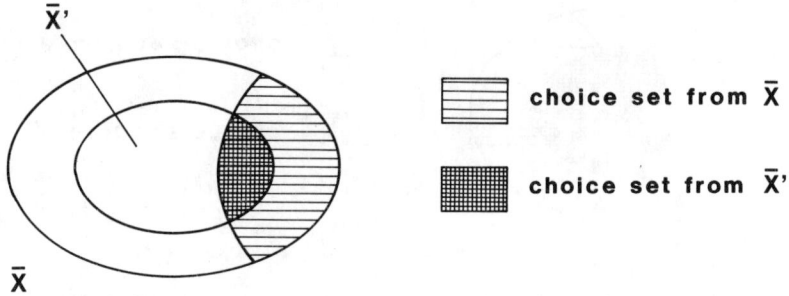

Fig. 8.1. WARP not violated

To show that a procedure violates WARP is in a way much simpler. All one has to do is to find one situation which is not like that in Figure 8.1. An example is Figure 8.2. where one of WARP's conditions, viz. requirement (i), is violated. One can see that not all winners in X belong to the set of winners in X'. This shows that requirement (i), and hence WARP is violated. In Figure 8.3. the other condition of WARP, viz. requirement (ii), is violated. In that figure the choice set of X' is larger than the intersection of the choice sets of X and X'. Thus, there are alternatives which are among the best ones in X' but are not included in the choice set of X even though some of the best alternatives of X' are also among the best in X.

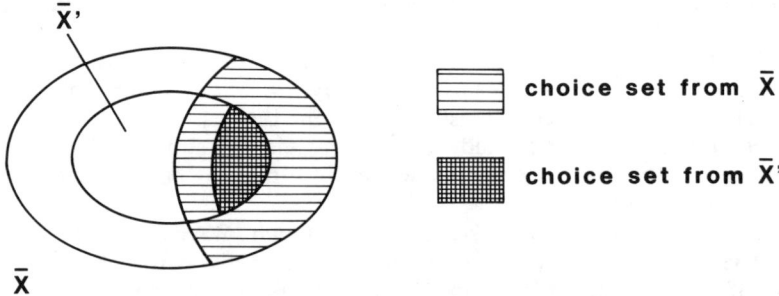

Fig. 8.2. Heritage violated, hence WARP and PI violated

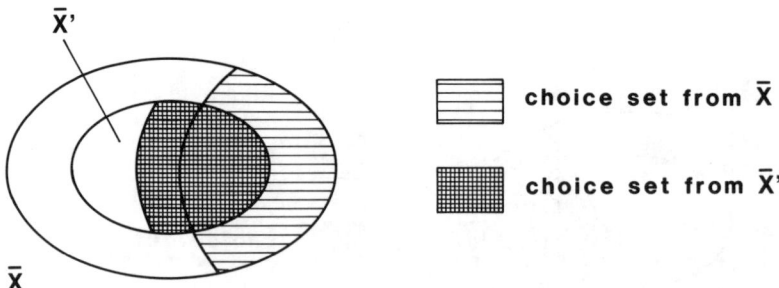

Fig. 8.3. WARP violated because (ii) violated, PI not violated if ▤ is also the choice set from F($\bar{X}$') U $\bar{X}$-$\bar{X}$'

The classic example of a violation of WARP is due to Sen (1970, p. 17): if the world champion in some game is a Pakistani, then all champions of Pakistan must be world champions in the same game. In other words, if two Pakistanis tie in the competition for the championship of Pakistan and the world champion is a Pakistani, then both of them must be world champions. It would, indeed, be plausible to require that if the world champion is a Pakistani, he must also be the champion of Pakistan, but less plausible to require that both of the tied players in the championship contest of Pakistan should be world champions assuming that one of them is the world champion. One of them could e.g. beat all non-Pakistani contestants, while the other might perhaps not be able to do so.

It is perhaps worth pointing out that requirements (i) and (ii) have been given somewhat varying names in the literature. Condition (i) is sometimes called property α (Sen, 1970, p. 17) or heritage (Aizerman and Malishevski, 1981, p. 1033), whereas condition (ii) is called property β. WARP itself has also been called the strict heritage or the keeping constancy of residual choice (Aizerman and Malishevski, 1981, p. 1033). We shall, however, use the concept of WARP because it seems to have been more widely adopted. On the other hand, when we shall need the intuitively more plausible part of WARP, viz. condition (i), the term heritage will be adopted.

Another property that relates to choice set invariance and also belongs to intra-profile properties is path-independence (PI) introduced by Plott (1973). It is the requirement that social choice should not depend on the partitionings of the alternative set. More specifically, we should end up with the same set of winners by using either of the following two methods of choosing:

1. applying the procedure to the entire set of alternatives given the preference profile, or
2. choosing first the winners, say set A, by applying the procedure to any subset X' of X given the restriction of the preference profile to that subset and then choosing the "final" winners from the set consisting of the winners just named and the rest of alternatives (X-X') using the same preference profile restricted to A ∪ (X-X'). In other words, PI requires that if one first picks the winners in a proper subset of X and then considers these winners together with the rest of the alternatives, one should come up with the same choice as when one considers the entire alternative set.

Obviously, when PI is satisfied, i.e. one always ends up with the same set of best alternatives no matter which of the two methods is resorted to, then one does not have to consider all the alternatives simultaneously, but can focus on a "manageable" subset without affecting the final outcomes. More formally, PI means the following. F is path-independent iff for any partitioning X_1, X_2 of X (i.e. $X_1 \cup X_2 = X$ and $X_1 \cap X_2 = \emptyset$) $F(X,R) = F(F(X_1,R) \cup X_2,R)$.

Prima facie PI would also seem to capture much of the intuitive meaning of "socially best alternative". One would not wish that the alternative(s) which the procedure specifies as best is (are) dependent on the order in which the alternatives are brought into consideration. The plausibility of PI is enhanced by the result of Aizerman and Malishevski (1981, pp. 1034-1035) according to which PI is actually the carrier of two properties viz. heritage or condition (i) of WARP and independence of rejecting the outcast variants (alternatives) (IRO). The latter property says the following: if $F(X,R) \subseteq X' \subseteq X$, then $F(X',R) = F(X,R)$. More informally, the condition requires that if all the winners in a large set (i.e. X) belong to a subset (X'), then these same winners should remain winners when the same procedure is applied to the subset only. If this requirement is applied to the previous example, it states that if all the world-champions in a given sport are Pakistani, then they must all be also the champions of Pakistan and all champions of Pakistan must be world-champions.

Prima facie IRO looks very much like the famous condition of the independence of irrelevant alternatives featured in the Arrow impossibility theorem. Strictly speaking IRO is, however, an intra-profile property while the independence of irrelevant alternatives (IIA) is an inter-profile one. What the latter says is that if two profiles are identical over a subset of alternatives, then the ensuing social choices from this subset should also be the same provided that the choice is restricted to this subset. However, IRO could also be viewed from an inter-profile viewpoint. Thus, $F(X,R)$ is a choice set from X when a group with preference profile R is considering X and $F(X',R)$ is a choice set when a different group with profile R as well is considering a subset of X. In the latter case R has to be interpreted as the restriction of the former R to the subset X'.

The difference between IRO and IIA can now be seen. IRO states something in such cases only where $F(X,R) \subseteq X'$, whereas IIA says something for all such situations in which two groups have identical profiles over a subset of X. If IRO is violated, then IIA is also violated. The converse, however, is not true. It may happen that in that particular subset X' to which $F(X,R)$ happens to be included IIA and IRO are both satisfied, but outside it there is at least one such subset X" in which IIA is not satisfied. The following example illustrates the difference.

FIFTH PROBLEM

Example 8.1.

Consider the following two preference profiles:

Profile 1:
person 1	person 2	person 3
a	b	c
b	c	b
c	a	a

Profile 2:
person 4	person 5	person 6
a	e	d
d	d	e
e	b	c
b	c	b
c	a	a

Supposing that the Borda count is used, IRO is satisfied in both profiles, i.e. in profile 1 (profile 2, respectively) b (d), the Borda-winner, is the winner in each subset to which it belongs. Looking now at the inter-profile property IIA, we observe that the profiles are identical over {a,c} and yet in the former c and in the latter a gets the higher Borda score. Hence in this example IRO is not violated while IIA is.

Figure 8.4. and Figure 8.5. illustrate the properties of IRO in the light of the Venn diagrams. In the former figure IRO is not violated as the choice set from X coincides precisely with the winners of X'. If this holds for every subset of X which contains all the alternatives chosen from X, then the procedure satisfies IRO.

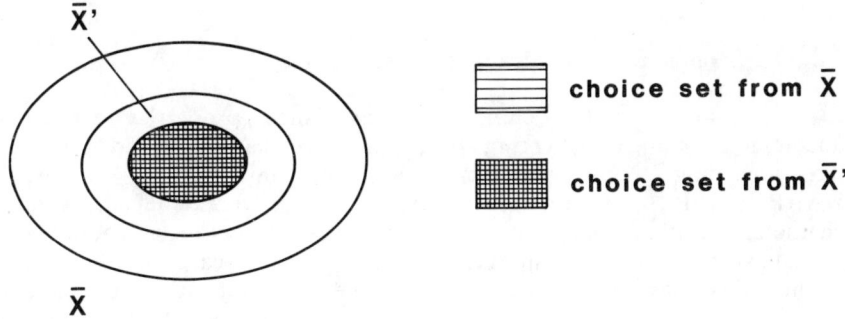

Fig. 8.4. IRO not violated

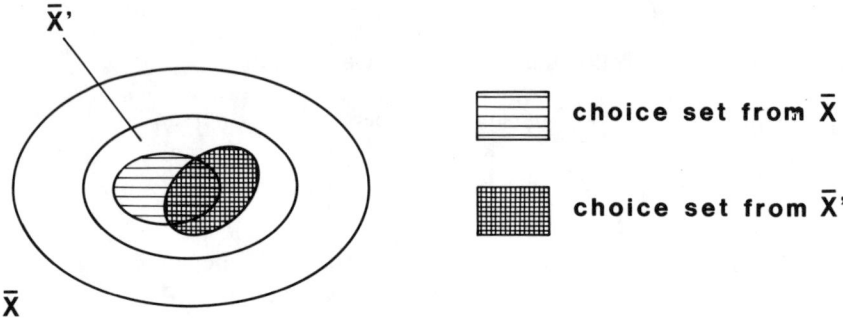

Fig. 8.5. IRO violated

In Figure 8.5. the choice set from X is entirely within X', but the choice sets from X and X' are not precisely the same. Hence IRO is violated.

We notice then that WARP and PI are rather similar properties of choice procedures. Indeed, they share one characteristic, viz. heritage. In other words, a violation of the heritage property means automatically the violation of both WARP and PI. But what then are the differences between these properties? An obvious answer would be to point to the fact that WARP is the carrier of (ii) in addition to heritage whereas the additional property of PI is IRO. It turns out, however, that WARP is a more stringent requirement than PI (see Aizerman and Malishevski, 1981, pp. 1033-1035). In other words, whenever a procedure satisfies WARP it eo ipso satisfies PI, but not vice versa.

8.2. Performances with respect to consistency

It turns out that the choice set invariance properties are very uncommon among the voting procedures we have touched upon in the preceding. Most uncommon are the intra-profile properties WARP and PI, but also consistency is rather rare. Indeed, it only characterizes the positional procedures, i.e. the Borda count, approval voting and the plurality method. In the case of the Borda count, this has been known since the axiomatization by Young (1974) was introduced. One of the axioms of the Borda count is consistency. Young also points out that the plurality procedure is consistent. Obviously the approval voting is consistent as well because if a subset of X is chosen by two distinct groups of

FIFTH PROBLEM

voters by the approval voting, it means that each e
subset gets the same number of approvals in each
number is maximal in each group. When the groups
then clearly the total number of approval votes is ma...
each element of the subset as the total is obtained by adding the
approval votes in groups.

All other procedures are inconsistent. Instead of showing this for each procedure separately, we utilize the result achieved by Young (1975) according to which all procedures that satisfy anonymity, neutrality and consistency are incompatible with the Condorcet winning criterion. The proof of Young's result is quite extensive and will not be reproduced here. The fact that anonymity, neutrality and consistency imply incompatibility with the Condorcet winning criterion ensues from the theorem according to which a social choice function is a scoring function iff it is anonymous, neutral and consistent. Scoring functions, in turn, determine the social ordering of alternatives on the basis of their total scores obtained from individual preference orderings e.g. in the way the Borda scores are computed. As these scoring functions are not compatible with the Condorcet winning criterion, it follows that no anonymous, neutral and consistent choice function is.

As we have already used the Condorcet winning criterion in the evaluation of the procedures, we can utilize this result because neutrality and anonymity are satisfied by all the procedures we are considering except by the amendment procedure in normal legislative settings where the so-called status quo alternative is given a special role, viz. it is always present in the final pairwise comparison. Neutrality requires that the relabelling of alternatives does not affect the social choice. However, we have not analyzed the amendment procedure in this way here, but considered it as a neutral procedure. This interpretation is natural in general contexts and also in legislative ones whenever no obvious status quo alternative is at hand.

All our procedures are also anonymous which means that the relabelling of the voters does not affect the social choice. In practice some non-anonymous procedures can be found; e.g. the procedure, where in the case of a tie the chairman's preference is decisive, is obviously non-anonymous. However, no such tie-breaking devices have been included in any of our procedures.

Young's result is thus immediately applicable. We can conclude that all such methods that satisfy the Condorcet winning criterion are inconsistent. This rules out all the binary procedures from the class of consistent ones. Similarly Black's and Nanson's methods turn out to be inconsistent. This leaves us to show that the plurality runoff as well as Hare's and Coombs' elimination methods are inconsistent. Let us start with the plurality runoff (see Nurmi, 1983a, p. 203).

Example 8.2. $X = \{x,y,z\}$
$N_1 = \{1, \ldots, 6\}$
$N_2 = \{7, \ldots, 11\}$

N_1's preference profile R:

voter 1	voters 2 and 3	voters 4-6
z	x	y
x	y	z
y	z	x

N_2's preference profile R':

voter 7	voters 8 and 9	voters 10 and 11
y	z	x
x	y	y
z	x	z

Now in N_1 no alternative has more than 50 % of the first ranks. Hence the runoff takes place between x and y. The result is a tie between x and y. In N_2 on the other hand, the runoff contestants are x and z with x the final winner. If the decision making body is $N = N_1 \cup N_2$ the runoff is between x and y. In this contest y wins. Hence, we have $F(X,R) \cap F(X,R') = \{x\}$, but $F(X,S) = \{y\}$ where S is the preference profile of N. This shows that consistency is violated.

The failure of the Hare system on the consistency criterion is by now well-known (see e.g. Doron, 1979; Fishburn and Brams, 1983). Consider the following example.

Example 8.3. $X = \{x,y,z,w\}$
$N_1 = \{1, \ldots, 12\}$
$N_2 = \{13, \ldots, 21\}$

N_1's preference profile R:

voters 1-4	voters 5-7	voters 8-12
x	y	z
y	x	y
w	w	w
z	z	x

FIFTH PROBLEM

N_2's preference profile R':

voters 13-15	voters 16-20	voter 21
y	x	w
w	w	y
z	z	z
x	y	x

$F(X,R) = \{x\}$ as y is first eliminated whereupon x gets more than 50 % of the first ranks. $F(X,R') = \{x\}$ as well because x has more than 50 % of the first ranks to start with. However, when the groups are combined we get $F(X,S) = \{y\}$ after w and z have been eliminated. Thus, consistency is violated.

Coombs' procedure is also incompatible with consistency as can be seen from the following example.

Example 8.4. $X = \{x,y,z,w\}$
$N_1 = \{1, \ldots, 9\}$
$N_2 = \{10, \ldots, 19\}$

N_1's preference profile R:

voters 1-4	voters 5-7	voters 8 and 9
x	y	z
y	z	x
w	w	y
z	x	w

N_2's preference profile R':

voters 10-12	voters 13-16	voter 17	voter 18	voter 19
y	x	y	z	x
w	w	x	x	y
z	z	z	y	z
x	y	w	w	w

Now $F(X,R) = \{x\}$ after z has been eliminated. $F(X,R') = \{x\}$ as well, this time after the elimination of y. However, $F(X,S) = \{y\}$ after the elimination of x. Thus Coombs' method is also inconsistent.

We can conclude that consistency is indeed a relatively rare property among the voting procedures. Only the positional methods possess it. It could even be argued that the property is not defined for approval voting as this method operates on ballots rather than preferences. Hence we cannot infer the winners directly from the preference data. However, in the assessment of consistency it is very plausible to assume that given the preferences of the subgroups the approved alternatives also remain the same when the groups are combined. This assumption is implicitly made above.

Another possibility in the assessment of the approval voting is to speak of ballot consistency instead of consistency simpliciter (Richelson, n.d.). This requirement can be expressed as follows. Suppose that $F(X,B_1) \cap F(X,B_2) \neq \emptyset$, where B_1 and B_2 are the ballots of two distinct groups of voters. Then F is ballot consistent iff $F(X,B_1) \cap F(X,B_2) = F(X,B)$ for all ballot sets B_1 and B_2. Here B denotes the ballot set of the combined group. Obviously the approval voting satisfies ballot consistency and this requirement is precisely the same as the assumption made throughout, viz. that each voter votes according to his preferences, in conjunction with the additional assumption that all voters in N_1 and N_2 approve exactly the same alternatives when voting in separate groups as when voting in N.

8.3. Performances with respect to WARP and PI

As was pointed out above, WARP is the more stringent of the two main intra-profile properties discussed in the preceding. This means that if a procedure satisfies WARP it necessarily also satisfies PI, while the converse is not true. On the other hand, a failure in PI implies by necessity a failure in WARP. Moreover, WARP and PI share one property, viz. heritage. Hence a violation of heritage is a sufficient (though not necessary) condition of a violation of both WARP and PI.

It turns out that a large majority of the voting procedures considered above are incompatible with heritage. The Condorcet paradox discussed in section 4.1. (Example 4.1.) shows that the amendment procedure violates heritage because a is among the winners in the set {a,b,c} but not in the set {a,c}. The same example can be used to demonstrate that Copeland's, Dodgson's and Schwartz' procedures along with the maximin method all fail on heritage as there is a three-way tie in the set {a,b,c} for each of these procedures. However, for each of the two-element subsets of X there is a (trivial) Condorcet winner which will be chosen by each of the methods. Hence in these two-element subsets there is an element that is among the winners in X but not in one of the subsets of X to which it belongs. By a similar reasoning it can be shown that Nanson's method fails on heritage as well.

The above example (Example 4.1.) can also be used to show that two positional methods, the plurality and Borda methods, fail on heritage and hence on WARP and PI. There is obviously a three-way tie in {a,b,c} when either of these procedures is utilized. However, when {a,c} is considered c will be the winner in either of these procedures assuming in the case of plurality voting that person 2 votes for his preferred alternative of the two, viz. c.

To see that the plurality runoff also violates heritage, consider the following example (Nurmi, 1983a, p. 199).

Example 8.5. $X = \{a,b,c,d\}$ $N = \{1, \ldots, 5\}$

voter 1	voters 2 and 3	voters 4 and 5
a	c	b
d	a	d
b	d	c
c	b	a

Obviously the plurality runoff winner in X is b. However, it is not the winner in a subset to which it belongs, viz. {a,b,d}. The winner in this subset is a as it has more than 50 % of the first ranks in the reduced alternative set.

Example 8.5. also shows the incompatibility of both Hare's and Coombs' procedures with heritage. After the elimination of d and a, the former specifies b as the winner in X. However, in the subset {a,b,d}, the Hare winner is a. Coombs' procedure, on the other hand, chooses d from X after the elimination of a and b. Considering again the subset {a,b,d}, the Coombs winner is not d, but a.

Of all our procedures then only the approval has not been shown to violate heritage. Indeed, supposing that the reduction of the choice set does not affect the "acceptability" of the alternatives, it can readily be inferred that the winners will remain winners in each subset they belong to under the approval voting. Moreover, if there is a tie between some alternatives in a subset of X, then if one of the tied alternatives, say x, is among the winners in X, so are the other ones because x cannot get more approvals than the others as all the voters who voted for x in the subset will - under our assumption - vote for it when X is considered. But the same holds for the rest of the tied alternatives. Hence if one of them gets the maximum number of approval votes in X, so will the others as the number of approvals is the same for x and the other alternatives which tie in the subset. Of course, it can happen that there are some new winners in X along with the ones that tied in the subset. It may even be the case that none of the tied ones will be among the winners in X. But these possibilities are not at all incompatible with WARP. Hence we can conclude that the approval voting satisfies WARP, and consequently PI.

8.4. The relevance of the criteria

The choice set invariance criteria capture some intuitively plausible properties of a good social choice procedure. Indeed, a failure to satisfy them would call into serious question the notion of the "best" alternative being chosen by the procedure. There are, however, differences in the nature of the above criteria. If consistency is violated, the subdivision of the voter set into subgroups has in some cases counter-intuitive effects on the outcomes. This feature can, therefore, be utilized to the benefit of the individual or group which is capable of making such subdivisions for practical purposes (e.g. the vote counting can be made more efficiently if subgroups are considered instead of the entire voter set). It is, inter alia, for this reason that the decision about where to draw the boundaries of electoral districts is of such great political importance. However, if a consistent voting procedure is utilized, then the boundary-drawing does not affect the electoral outcomes.

Using consistent procedures thus makes it possible to "decentralize" the decision making into several subgroups. The alternative set considered by each group is, however, the same. If WARP is satisfied, then another type of decentralization is possible, viz. the subdivision of the alternative set. WARP guarantees that a person controlling the agenda cannot, by introducing purely hypothetical or straw-man alternatives, affect the outcomes unless the latter ones are actually chosen. Similarly, PI is a property that excludes certain types of manipulations, viz. those involving the order in which the alternatives are brought to the consideration of the voting body. We shall return to these strategic properties shortly (9.4.; see also Nurmi, 1984b).

As was pointed out above, WARP is the strongest of the intra-profile choice set invariance properties. One of its "subproperties", heritage, is intuitively very compelling, while the other, property β, is less so. The reason is that it seems to be in the nature of what we mean by a winning alternative that it is a winner in all such alternative sets that are smaller than the original set, because it then has a smaller number of contestants that it already won in the original set. On the other hand, property β requires that the performance of a set of tied alternatives with respect to a larger set does not differentiate between them. This certainly is not conceptually related to our intuitions about winning as our previous example about the world- and Pakistani championship shows. It would seem then that one can well do without property β. But as our preceding discussion shows the main culprit in the almost universal incompatibility with WARP is that heritage is a very rare property.

CHAPTER 9

SIXTH PROBLEM: HOW TO ENCOURAGE THE
SINCERE REVELATION OF PREFERENCES

One of the main motivations for resorting to collective decision making is to elicit the preferences of the persons involved and aggregate these in a meaningful way into a collective choice. The very idea of individual preference based social choice rests on the assumption that the individual preferences used as inputs in the process of preference aggregation are the true preferences of the individuals. Indeed, the collective goods provision problems are sometimes "solved" by resorting to voting procedures. These problems stem from the fact that those optimality results that characterize private good economies with large numbers of buyers and sellers are not generalizable into economies dealing with public or collective goods (or "bads", for that matter). In particular, the result according to which the perfect market mechanism leads to Pareto optimal resource allocation when private goods are traded does not apply to public goods. In other words, while for the private goods individual utility maximizing behaviour leads to a situation where nobody's welfare can be increased without worsening at least one other individual's welfare, this is not the case for public goods (see Riker and Ordeshook, 1973, pp. 244-249). For the latter type of goods one may actually expect that there will be a Pareto sub-optimal amount of public goods available (Olson, 1965). When the collective goods provision problems are "solved" by resorting to voting procedures, the idea is to give the individuals an incentive to reveal their true preferences concerning the amount of good to be produced or bought by the collectivity. This incentive is lacking to some extent when the market mechanism is used. The voting "solution" is based on the idea of removing the free rider element from the situation (see e.g. Hardin, 1971). This latter element, in turn, consists of the fact that in a market situation it may not be in an individual's interest to reveal his true preference concerning the public good because by doing so he may be the sole payer of the costs involved, while everybody would be the beneficiary of the good.

SIXTH PROBLEM

The literature on the public goods provision and free rider problems is vast (see e.g. Hardin, 1982; Laver, 1981; Olson, 1965; Taylor, 1976). So is the literature on preference revelation mechanisms and incentive compatibility (see e.g. Feldman, 1980; Green and Laffont, 1979; Tideman and Tullock, 1976). Perhaps it should be pointed out that even though the voting procedures can sometimes be thought of as "solutions" to the collective goods provision problems, they are by no means the only devices that have been employed for this purpose. E.g. central governments often deal with collective goods provision and thus the link to individual preference revelation is of an indirect nature. Let us, however, look at the performance of voting procedures in collective goods provision.

In voting by secret ballot one would be more motivated to indicate his true preferences because if the level of provision were determined on the basis of the social choice, given the indicated preferences, and the costs were distributed among the participants regardless of their individual ballots, then the link between preference revelation and cost share would be removed. It turns out, however, that this reasoning does not carry us far enough, i.e. it still remains in the individual's interest to misrepresent his preferences at least in some contexts.

9.1. Manipulability

It is customary to say that a social choice function is manipulable by an individual, say i, in a situation which involves the alternative set X and the preference profile R, iff the outcome resulting from F when applied to set X and preference profile R is worse for i than the outcome that would ensue if all other things remained as previously but i's reported preference were different. In other words, F is manipulable by i precisely when there is a situation (i.e. a (X,R) pair) in which i can benefit from not revealing his true preference order - that is, the preference included in R - but another preference order. More formally, we say that the social choice function F is manipulable by i in situation (X,R) iff

$F(X,R')$ P_i $F(X,R)$, i.e. i prefers strictly $F(X,R')$ to $F(X,R)$, where
$R = (R_1, \ldots, R_{i-1}, R_i, R_{i+1}, \ldots, R_n)$,
$R' = (R_1, \ldots, R_{i-1}, R'_i, R_{i+1}, \ldots, R_n)$ and $R'_i \neq R_i$.

In other words, i strictly prefers the outcome resulting from the social choice function, given the preference profile R', to that resulting from the preference profile R where the only difference between these two profiles is i's own preference. Supposing that R_i is i's true or sincere preference, R'_i is then the insincere preference of i.

We say then that the social choice function or the procedure that realizes it is manipulable by i in situation (X,R) exactly when i can benefit from reporting that R'_i instead of R_i is his preference relation. The social choice function is (individually) manipulable iff there is a situation such that some individual can benefit from insincere preference revelation, i.e. there is an individual i such that i can manipulate F in that situation. If F is such that there is no situation in which some individual could benefit from insincere preference revelation, then F is strategy-proof, i.e. the strategic misrepresentation of preferences will never benefit any voter. There are other forms of manipulability as well, e.g. related to agenda-control, but in this section we shall deal with manipulability in the strict sense in which it refers to preference misrepresentation. Other forms are discussed later on.

In the mid-70's Gibbard (1973) and Satterthwaite (1975) published independently a result concerning the manipulability of social choice functions that produced an extensive study of this aspect of choice procedures (see e.g. Pattanaik, 1978; Gärdenfors, 1976; Kelly, 1978, pp. 65-82; Barbera, 1977). The result can be considered an impossibility result in a similar way as that of Arrow. To state the result we need some new definitions.

First, a social choice function F is said to be universal iff it is defined for all situations. Second, F is nontrivial iff for each element x of X, there is a situation such that F results in x in that situation. Third, F is single-valued iff F(X,R) is always a singleton set, i.e. F specifies a unique winner in all situations. Fourth, F is dictatorial iff there is an individual i such that F(X,R) is always the same as i's most preferred alternative (see e.g. Feldman, 1980, pp. 202-212). The Gibbard-Satterthwaite theorem says that all universal, nontrivial and single-valued social choice functions are either manipulable or dictatorial. Formulated as an impossibility result the theorem states that no social choice function satisfies universality, nontriviality, single-valuedness, nonmanipulability and nondictatorship.

The result says that if we have a social choice procedure that satisfies the intuitively plausible properties of universality, nontriviality, single-valuedness and nondictatorship, then we should not be astonished to find a situation in which it is at least in one individual's interest not to reveal his true preference order but to act as if the latter were a different one. Hence we cannot count on the fact that the voting procedure used would elicit the true preferences of the individuals in every situation. Of course, we must first determine whether the voting procedure we propose for use is such that the above properties are satisfied. As will become evident shortly, we shall usually have to make do without one of the above properties, plausible as they seem at first sight.

As an individual can be considered as a coalition consisting of one member only, it follows that all social choice functions that are manipulable by individuals are also manipulable by coalitions. However, the converse is not necessarily true, i.e. there can be situations which are manipulable by coalitions of several individuals but not by one individual.

But what is so bad in manipulability in the first place, one could ask. After all, for each individual there is usually at least some options that they would rank as highest if these were put on the agenda. So why bother fiddling with orders of alternatives that happen to be on the agenda? One answer is to point to the fact that if the procedure is manipulable, then one cannot with any certainty say very much about its other (profile-dependent) properties, i.e. ones that are defined with respect to fixed ("true") preference orders. In other words, if a procedure has some property when the preferences are assumed to be fixed, then one cannot be sure that this property will ever materialize in reality if the fixed preferences involve incentives not to reveal them. Stated yet in another way, if a procedure P has property T, this means that e.g. when the preference profile is R the choice set has some specific characteristics (e.g. some outcomes are excluded). But if, on the other hand, whenever the voters have true preferences as in R they act as if they had preference profile R' which differs from R, then this means that $F(X,R)$, that is, the "theoretical" choice differs from $F(X,R')$ which is the outcome after manipulation. Thus F does not in practice have the properties it has when viewed theoretically.

The second answer could be that often the very act of going to the people has the sole rationale of finding out the true preferences of people involved in voting. For example, it makes no sense to conduct a referendum on an issue involving more than two response alternatives if there is a presumption that people do not reveal their true preferences (assuming that they have ones). Neither of these answers can play down the importance of agenda-control, but is intended to point to the relevance of preference misrepresentation for the rationale of voting as an institution.

9.2. Performance with respect to manipulability

Prima facie the Gibbard-Satterthwaite theorem would seem to undermine the very rationale of the voting procedures, viz. their use in eliciting the true preferences of voters concerning the alternatives at hand. It would then seem that voting procedures are not, after all, capable of resolving the problem of optimal public goods provision. On closer inspection, however, these conclusions turn out to be unwarranted. The reason for this is simply that the Gibbard-Satterthwaite theorem does not apply to the voting procedures we have been discussing. None of them is single-valued and, hence, the theorem does not say anything about these procedures.

There is, however, another theorem which is directly applicable to several of our voting procedures, viz. that proven by Gärdenfors (1976). It states that all anonymous and neutral procedures satisfying the Condorcet winner criterion are manipulable by individuals. Instead of presenting the proof in detail we shall just sketch it. Consider first the three alternative, three voter case and the following two profiles:

	person 1	person 2	person 3
Profile 1:	z	x,y	x
	y		z
	x	z	y
Profile 2:	z	y	x
	y	x	z
	x	z	y

Gärdenfors considers all possible choice sets from profile 1 and shows that for each choice one can find a preference profile such that the choice function used is manipulable either at profile 2 or at some other profile. Without covering all cases, let us consider the one in which the choice function used specifies x and y as choices from profile 1. The anonymity and neutrality of choice functions imply that the choice set from profile 2 must be {x,y,z} as this is the Condorcet paradox preference profile in which all alternatives are once ranked first, once second and once third. In other words, there is no possibility of making any difference between them without violating either neutrality or anonymity. But under these assumptions profile 2 is manipulable by person 2 because by acting as if his preference were that of person 2 in profile 1, he can bring about the outcome {x,y} which he prefers to {x,y,z} tie.

To take another example, suppose that the choice from profile 1 were z. Then person 2 could by misrepresenting his true (i.e. profile 1) preferences bring about the following profile:

Profile 3:   person 1   person 2   person 3
                     z   x   x
                     y   y   z
                     x   z   y

In profile 3, x is the Condorcet winner and must thus be chosen if the procedure satisfies the Condorcet winning criterion. But then obviously the procedure is manipulable by person 2 at profile 1 because by acting as if his preferences were the same as in profile 3, he can bring about the outcome x which he certainly prefers to z. Indeed, he prefers x to any outcome which includes z.

Similar examples can be presented for each possible choice from profile 1. The conclusion is then that for three alternatives and three voters all social choice functions satisfying the Condorcet winner criterion, neutrality and anonymity are manipulable. The next step in Gärdenfors' proof is to generalize the result to larger voter and alternative sets. This can be done by simply adding "dummy" alternatives in fixed orders below x, y and z in every individuals' preference orders. Similarly dummy voters who are indifferent between x, y and z can be used to construct larger voter sets.

One should perhaps add that Gärdenfors' theorem does *not* say that the procedures satisfying the Condorcet winning criterion, anonymity and neutrality are manipulable whenever the preference profiles in question have a Condorcet winner. It does not say anything at all about specific preference profiles. What it says is that if a procedure has the above three properties, then there are profiles under which an individual may benefit from acting as if his preferences were different from what they really are. The proof of this theorem makes extensive use of specific preference profiles, as we have just seen, but the result does not say under which profiles the individuals can manipulate the various procedures. In fact, there are results suggesting that if *all* voters act strategically, then the profiles in which there is a Condorcet winner are particularly immune to manipulation if certain procedures are used (see McKelvey and Niemi, 1978; Miller, 1980; Banks, 1985). We shall return to this in section 12.1. However, Gärdenfors' theorem does not deal with specific profiles.

Obviously all the procedures we have considered are anonymous and all but one - viz. the amendment procedure - neutral (see Riker, 1982, p. 101). The fact that the amendment procedure is non-neutral is due to the special status given to the status quo alternative. If in this procedure a relabelling occurs so that some alternative which was formerly regarded as an amendment is now the status quo alternative, then this change means that it is now necessarily present in the final pairwise comparison, whereas the former status quo alternative is perhaps in the very beginning the voting sequence. Obviously, the relabelling then affects the voting agenda and consequently there is no assurance that the social choice remains the same after relabelling. However, the following example shows that the amendment procedure also falls within the realm of manipulable social choice procedures.

Example 9.1. The Condorcet paradox. $|N| = 3$, $X = \{a,b,c\}$

voter 1	voter 2	voter 3
a	b	c
b	c	a
c	a	b

Suppose that the agenda is:

1. a vs. b, and
2. the winner of 1. vs. c.

Hence c is the status quo alternative. If every voter votes according to his true preferences in both stages, the winner is c. (Note that a relabelling of alternatives so that a would be the status quo alternative would make a the winner.) Suppose now that voter 1 does not indicate his true preference relation, but the following: bac. If no other changes occur, then b defeats a in 1. and becomes the over-all winner having also beaten c in 2. Obviously the outcome b is better from voter 1's view-point than c. Hence 1 has benefited from an insincere revelation of preferences. Thus the procedure is manipulable.

Gärdenfors' theorem and Example 9.1. allow us thus to conclude that all our binary procedures plus Black's and Nanson's methods are vulnerable to insincere preference revelation. But so are the positional procedures even though they do not fall within the realm of applicability of Gärdenfors' theorem because of their failure on the Condorcet winning criterion. Consider the following examples.

SIXTH PROBLEM

Example 9.2. |N| = 5, X = {a,b,c}

voters 1,2	voters 3,4	voter 5
a	b	c
b	c	a
c	a	b

If the plurality method is used, the result is a tie between a and b provided that each voter votes for his most-preferred alternative. Now under a rather mild assumption concerning voter preferences voter 5 can benefit from an insincere preference revelation if he votes for a instead of c since by so doing he switches the a-b tie into the victory of a. It is quite reasonable to assume that a voter prefers a to the a-b tie whenever he prefers a to b.

Actually this assumption is a less stringent requirement than what is called the monotonicity in prizes assumption (see e.g. Harsanyi, 1977, p. 33). The latter requires the following: if a is preferred to b, then the lottery (a, p; c, 1-p) is preferred to the lottery (b, p; c, 1-p) for any p in the [0,1] interval. Here (a, p; c, 1-p) denotes the lottery in which the probability of prize a is p and the probability of prize c is 1-p. Mutatis mutandis the same interpretation holds for (b, p; c, 1-p). In other words, if there are only three prizes in two lotteries, one of which is common in both lotteries and occurs in them with the same probability, then the lottery in which the other prize is higher is preferred to the one in which it is lower, assuming that these unequal prizes occur with same probability in both lotteries. The assumption that a is preferred to the a-b tie whenever a is preferred to b is obviously weaker than the monotonicity in prizes assumption.

The following example shows that the Borda count is also manipulable.

Example 9.3. |N| = 3, X = {a,b,c,d}

voter 1	voter 2	voter 3
a	b	d
b	d	a
c	c	b
d	a	c

The Borda scores of the alternatives are a (5), b (6), c (2), d (5). So the winner is b. If now voter 3 misrepresents his preference as acdb, the new Borda scores are: a (6), b (5), c (4), d (3). Now the winner is a, which is preferred to the former winner b by voter 3.

Hence we conclude that the Borda count also is manipulable. To show that the approval voting also is manipulable, we could refer to Example 9.2. because the plurality method is a special case of the approval voting, viz. each voter approves one alternative only. Slightly more in line with the spirit of the approval voting is the following example.

Example 9.4. $|N| = 3$, $X = \{a,b,c\}$

voter 1	voter 2	voter 3
a	b	c
b	a	a
c	c	b

Suppose now that voters 1 and 2 approve their two most preferred alternatives, while voter 3 approves c only. The result is then an a-b tie. Now by misrepresenting his approval so as to give a vote for a only (or for both a and c which amounts to the same result) voter 3 can bring about the victory of a which he under our previous assumption prefers to the a-b tie. Hence also the approval voting is manipulable.

Let us now turn to the multi-stage procedures and start with the plurality runoff. The following example which is a slight variation of 9.4. shows that this method also is manipulable.

Example 9.5. $|N| = 5$, $X = \{a,b,c\}$

voter 1	voters 2, 3	voters 4, 5
a	b	c
b	a	a
c	c	b

Now in this case the runoff takes place between b and c and the winner is b under the assumption of a sincere preference revelation. Suppose now that voter 4 votes for a instead of c in the first round. The runoff will then take place between a and b in which contest a wins. The outcome a is obviously preferred to b and, thus, the misrepresentation of preferences benefits voter 4 showing that the method is manipulable.

The two remaining procedures, Hare's and Coombs', are also manipulable as can be seen from the following example and the preceding one.

SIXTH PROBLEM

Example 9.6. |N| = 12, X = {a,b,c,d}

voters 1-4	voters 5, 6	voters 7-10	voter 11	voter 12
a	c	d	d	a
b	d	b	b	b
c	b	a	c	d
d	a	c	a	c

Since no alternative has more than 50 % of the first ranks, the alternative with the largest number of last ranks, c, is eliminated when Coombs' procedure is used. After the elimination d has more than 50 % of the first ranks, and is therefore chosen as the Coombs winner. Suppose now that voter 12 misrepresents his preference as abcd. Then d has the largest number of last ranks and is eliminated. After that we need another elimination, viz. that of c to come up with the winner which is now b. Clearly the preference misrepresentation has benefited voter 12 as he prefers b to d.

Example 9.5. shows that Hare's method is also manipulable. In that example b becomes the Hare winner after the elimination of a assuming a true preference revelation. Suppose now that voter 4 indicates the preference acb instead of cab. The Hare-winner is now a. Obviously voter 4 prefers a to b which then shows that Hare's procedure is manipulable.

Manipulability in the sense of the Gibbard-Satterthwaite theorem thus characterizes all the procedures we have considered above. It is, however, arguable that this sense is not necessarily relevant for the purpose of finding good preference aggregation procedures. The reason for this statement is that to regard a procedure as manipulable one only needs to show that there is a situation in which one individual benefits from not revealing his true preference. It is not required that the individual knows of this in the sense of knowing that the situation in question has materialized. Moreover, if the individual knows that the situation in which he would benefit from preference misrepresentation has materialized in terms of the individuals' true preferences, he may refrain from misrepresentation because it may occur to him that others may have incentives for misrepresentation as well. He may well come to the conclusion that in view of the likely misrepresentations of others his best action would be to reveal his true preferences. But, of course, he may also conclude that some new way of preference misrepresentation would give him the largest benefit. However, all these calculations are based on the anticipation of others' behaviour and the situation is thus likely to be

unstable in the sense that there are no single best ways of preference revelation.
   Another somewhat questionable assumption underlying this definition of manipulability is that an individual could think of changing his preference without the others' changing theirs. It would seem that the requirements concerning what an actor knows about the preferences of the others are simply too extensive to be of much practical importance. What is therefore called for is a more realistic definition of manipulability, one which would take into account that aspect in the above notion which seems pertinent, viz. the fact that manipulability relates to the stability of a voting procedure, without at the same time making the notion too stringent in terms of the informational requirements of the actors.

9.3. The difficulty of manipulation

The importance of manipulability in social choice theory is due to the fact that whenever a procedure is manipulable or vulnerable to the misrepresentation of individual preferences, there is no assurance that it satisfies the other possibly favourable properties that are based on a sincere preference revelation. Manipulability is in a sense a higher order property as it, to some extent, determines the lower level properties like the Condorcet criteria, Pareto optimality etc. which were discussed above under the assumption of a sincere preference revelation. But the fact that a procedure is manipulable does not per set make it likely that the individuals will necessarily misrepresent their preferences as it may be the case that they do not know how to benefit from such behaviour. For instance in the case in which a collective decision making body consists of persons who are total strangers to each other and a secret ballot is taken, there is simply no way any person could benefit from a misrepresentation of his preferences.
   On the other hand, one can easily imagine situations in which an individual can benefit from a misrepresentation of preferences even though he does not know the entire preference profile of the other voters. A case in point is the plurality voting where a successful manipulation does not require any other knowledge of the preferences of other voters than the distribution of first preferences. The following example makes this point evident.

SIXTH PROBLEM

Example 9.7. |N| = 5, X = {a,b,c,d}

voters 1, 2	voters 3, 4	voter 5
a	b	d
.	.	a
.	.	c
.	.	b

Thus voter 5 does not know more about the others' preferences than the distribution of the highest ranked alternatives. Such information if often obtainable e.g. from opinion polls. Given this information voter 5 can misrepresent his preferences with beneficial results by voting for a instead of d, his most preferred alternative. If voter 5 votes sincerely, i.e. for d, the result is a tie between a and his least preferred alternative b provided that voters 1-4 vote according to their true preferences. On the other hand, if voter 5 votes for a the result is the victory of a, his second-best alternative. Obviously, misrepresentation benefits voter 5.

Now the question arises as to whether it is possible to say something general about the difficulty of preference misrepresentation under various voting procedures. The following remarks purport to do just that by way of a manipulation hierarchy.

First of all we must define what is meant by the degree of vulnerability of various voting procedures to the strategic misrepresentation of preferences. This degree is intended to reflect the type of knowledge one *typically* needs in order to benefit from preference misrepresentation. The more detailed the knowledge of the preference profile one needs, the less vulnerable is the procedure to misrepresentation. Note that the degree of vulnerability is not directly defined in terms of the sufficient condition for successful misrepresentation; the kind of knowledge one typically needs in order to manipulate successfully may not in itself guarantee success. Rather it provides incentives for such behaviour and - viewed from another angle - casts doubt on the possibility of finding out the true preferences of voters.

More plausibly the degree of vulnerability reflects the necessary condition of a successful preference misrepresentation. It must, however, be kept in mind that a person may misrepresent his preference and thereby come up with a better outcome than under a truthful preference revelation, but he does not have to know anything at all about the preference profile. When this happens, success is due to sheer luck. So if one wants to define the degree of vulnerability in terms of the necessary amount of knowledge concerning the preference profile, then the case of a lucky manipulator has to excluded.

Of the procedures considered so far the most vulnerable to the strategic misrepresentation of preferences would seem to be the plurality procedure because one typically needs no other knowledge for a successful preference misrepresentation than the distribution of the first preferences of voters. Example 9.7. demonstrates this. It is noteworthy that the plurality runoff is not similarly vulnerable, i.e. a successful preference misrepresentation typically requires more knowledge of the preference profile than the distribution of the first preferences. For instance in Example 9.7. voter 5 would not benefit from voting a instead of d in the first round as the runoff - if one would ensue - would take place between a and b and he could thus vote sincerely in both rounds and end up with a as winner nonetheless. The lesser degree of vulnerability of the plurality runoff system can furthermore be illustrated by the following example.

Example 9.8. $|N| = 11$, $X = \{a,b,c,d\}$

5 voters	2 voters	2 voters	2 voters
a	b	c	d
b	.	.	.
c	.	.	.
d	.	.	.

Prima facie it would seem that one of the 5 voters with preference order abcd would benefit from voting for some other candidate than a on the first round because he would then decide which alternative would confront a in the second round. But if no other information than that given above is offered about the preference profile, the case for preference misrepresentation is ill-founded because the success of the manoeuvre depends on the distribution of the second preferences among the voters.

Suppose that one the 5 voters votes for b instead of a and the others vote according to their true preferences. Then the runoff takes place between a and b, certainly a reasonably good start for the 5-voter group. But it may well happen that in the second round b wins because all the voters, whose preferences are not entirely given, rank b higher than a. Suppose that in the absence of a clear winner among b, c and d in the first round, a's contestant in the second round - given a true preference revelation - is determined by a random mechanism of some kind. Suppose that the mechanism decides that d confronts a in the runoff contest. Then it is quite possible that a wins because it could be

SIXTH PROBLEM

ranked higher than d by more than 50 % of the voters. The fact that the five-member group has some "extra" votes in the first round does not, then, per se guarantee that its members benefit from preference misrepresentation. What is needed for that is a profounder knowledge of the preference profile.

The next level in the degree of vulnerability to the misrepresentation of preferences consists of systems which typically do not require more information about the preference profile than the distribution of approvals among the alternatives, i.e. how many voters regard any given alternative as acceptable. Obviously the approval voting belongs to this class of procedures.

Example 9.9. $|N| = 3$, $X = \{a,b,c,d\}$

voter 1	voter 2	voter 3
a, b	a, b	c
.	.	a
.	.	d
.	.	b

Thus all voter 3 knows about the preference profile of voters 1 and 2 is that they both approve a and b. Voter 3's preference order is cadb where only c is considered acceptable by him. This knowledge enables voter 3 to benefit from preference misrepresentation by voting for both c and a or just a. Thereby the victory of a ensues, an outcome that obviously is better for voter 3 than the tie between a and b which would otherwise be the result.

The third level in the misrepresentation hierarchy consists of systems which are based on the pairwise comparisons of alternatives so that in order to misrepresent one's preferences successfully one has to know the entire pairwise comparison matrix. The amendment and maximin procedures along with Copelands', Dodgson's and Schwartz' methods are all typically vulnerable to the misrepresentation of preferences given this type of knowledge. Consider the following examples.

Example 9.10. $|N| = 4$, $X = \{a,b,c,d\}$

voter 1   binary comparison matrix:

voter 1		a	b	c	d
b	a	-	3	2	3
c	b	1	-	3	3
a	c	2	1	-	3
d	d	1	1	1	-

One immediately observes that a would be chosen by each of the above mentioned binary procedures. Voter 1 could, however, change this outcome to his benefit by voting for c in the pairwise contest between b and c, i.e. by acting as if his preference were cbad. The following binary comparison matrix would then ensue:

	a	b	c	d
a	-	3	2	3
b	1	-	2	3
c	2	2	-	3
d	1	1	1	-

Now the choice set consists of a and c if either the maximin method or Schwartz' procedure is used. In other words, if either of these procedures is used voter 1 can benefit from preference misrepresentation given no other information than the binary comparison matrix.

To see that a similar conclusion applies to Copeland's and Dodgson's procedures as well, let us consider the following case.

Example 9.11. |N| = 5, X = {a,b,c,d}

voter 1   binary comparison matrix:

b
c
d
a

	a	b	c	d
a	-	3	2	3
b	2	-	3	2
c	3	2	-	3
d	2	3	2	-

The Copeland scores of a and c are both 1, while those of b and d are both -1. Hence the Copeland winners are a and c. Similarly a and c are the Dodgson winners as both can be rendered the Condorcet winner by only one preference change. This is not, however, necessarily the case if Fishburn's version of the Dodgson method is used (see 4.4.2.). This version needs a more profound knowledge of the preference profile for the preference misrepresentation to succeed.

Suppose now that voter 1 votes as if his preference order were cbda. Thereby a new pairwise comparison matrix results:

	a	b	c	d
a	-	3	2	3
b	2	-	2	2
c	3	3	-	3
d	2	3	2	-

Here c is now the Condorcet winner and will therefore be chosen by both Copeland's and Dodgson's methods. Obviously c is preferable to the a-c tie by voter 1 who therefore has benefited from the preference misrepresentation.

The fact that the binary methods are all vulnerable to preference misrepresentation of an equal difficulty is not surprising as the winning criterion of these methods is a binary one by definition. Hence all relevant information used by these methods is contained in the pairwise comparison matrix. We observe, however, that in concrete cases the utilization of the preference misrepresentation requires considerably more information than in the previous cases, i.e. in connection with the plurality and approval voting methods.

The fourth level in the preference misrepresentation hierarchy consists of procedures that are most difficult to manipulate in the sense that the informational requirements for a successful manipulation are most stringent: in typical cases one has to know the entire preference orderings of most - if not all - voters to benefit from the misrepresentation of one's own preference ordering. A procedure that most clearly represents systems at this level is Coombs' method. But typically the Hare system also requires a profound knowledge of the preference profile to enable a successful preference misrepresentation. One could also maintain that the plurality runoff method presupposes same kind of information as the two previous ones for a successful manipulation. In the absence of any meaningful estimate of the probability of various types of preference profiles, it is impossible to say whether Coombs' method typically requires more profound information about the preference profile than either the plurality runoff or the Hare system.

We have not yet placed the Borda count and the procedures that make use of it in the hierarchy. Prima facie the Borda count should be allocated to the same level of the hierarchy as the three above mentioned procedures because of its positional nature. However, as we have observed earlier (see 4.5.), the Borda count can also be implemented by using exclusively information about pairwise comparisons. Therefore, it should be placed on the third level of the hierarchy along with the binary procedures. Similarly the procedures based on the Borda count, i.e. Nanson's and

Black's, should be classified as belonging to the third level of the hierarchy.

The preference misrepresentation hierarchy can be summarized in the following table (Table 9.1.).

level	procedure	information required for successful misrepresentation
1st level	plurality	distribution of first preferences over alternatives
2nd level	approval voting	distribution of approvals over alternatives
3rd level	amendment, Copeland, Dodgson, Schwartz, maximin, Borda, Nanson, Black	binary comparison matrix
4th level	Coombs, Hare, plurality runoff	entire preference profile

Table 9.1. The preference misrepresentation hierarchy: the difficulty of benefiting from the misrepresentation of one's preferences.

A few comments on the hierarchy are now in order. Firstly, the procedures on the fourth level are often - although not typically - easily manipulable with a much less profound knowledge of the preference profile than the fourth level would imply. The cases in which an individual or a group of voters can render one alternative the winner by making its share of the first ranks exceed 50 % of the electorate whereas the alternative would not otherwise have this much support, are manipulable in principle with similar kind of information as the first level procedures require. This applies to all the procedures we have placed on the fourth level.

Secondly, it is to some extent a matter of taste whether the approval voting is considered more or less easily manipulable than the procedures on the third and fourth levels. The reason for this

SIXTH PROBLEM 125

ambiguity is that one could argue that even a full knowledge of the preference profile does not enable one to predict which alternative is the winner. In a sense then the manipulation of the approval voting requires even more knowledge of the preference profile, viz. knowledge of where the voters place their approval thresholds given their preference orders. On the other hand, the approval voting does not require the notion of preference ranking at all and, thus, it is more natural to regard the procedure as more easily manipulable than the systems on the third and fourth levels. We encounter here a similar ambiguity to the one we touched upon earlier in our discussion of ballot consistency (cf. 8.2.). Indeed, one form of manipulation in approval voting is changing one's approval threshold and not one's preference order. This type of manipulation is called sincere preference truncation which we shall discuss shortly (9.5.).

Thirdly, the most important conclusion to be drawn from the table is that there are marked differences between procedures and that therefore it is not of much consequence to discover that all of them are manipulable. This finding conceals many important differences between the procedures and has thus very little practical bearing.

9.4. Agenda-manipulability

Preference misrepresentation concerns the rationale of asking people to vote. If a procedure is vulnerable to preference misrepresentation in an obvious way, that is, if it requires very little information to make misrepresentation worthwhile, the very reason for resorting to voting in the first place collapses. Agenda-manipulation is a related problem and may, indeed, occur simultaneously with preference misrepresentation. Its crux, however, lies in the fact that sometimes the person or group controlling the agenda has a disproportionate power over the chosen outcomes vis-a-vis the other members of the voting body. It turns out that various procedures are vulnerable in different ways to agenda-manipulation at least insofar as the latter activity can be explicated in a precise fashion with the aid of a couple of concepts that we have already utilized in our preceding discussion.

What then do we mean by agenda-manipulation? Surely it can take on various forms in concrete settings, but two special forms may be worth looking into in some detail. Firstly, agenda-manipulation may mean that, by deciding in which order the alternatives are brought to the consideration of the decision making body, the person or group controlling the agenda can affect the chosen outcomes in a way that suits his or its purposes. This form has actually been the focus of theoretical interest for some time.

Perhaps the best-known result pertaining to this form of agenda-manipulation is due to Farquharson (1969) who suggested that if the amendment procedure is utilized and all voters vote according to their true preferences, the agenda-controller is well advised to place his own favourite last on the agenda. In other words, under these circumstances it is beneficial to have one's most-preferred alternative voted upon last or considered as the status quo alternative. In the case of three alternatives this result is quite straightforward; if another alternative which is not the status quo one wins, then it must be the Condorcet winner under the stated assumptions. However, the status quo alternative may become the winner of the process without being the Condorcet winner. All it needs to win is to defeat the alternative confronting it in the single pairwise comparison in which it is involved.

So if three alternatives are considered, it is easier for the status quo alternative to win than for the other two to do so. But obviously this argument can be generalized: for the status quo alternative to win, it needs to defeat just the contestant in the final pairwise comparison, while for any other alternative to win, it needs to defeat at least two alternatives one of which is necessarily the status quo alternative. So one can see why under sincere voting Farquharson's result holds.

This surely doesn't say anything about the voting procedures in general and of their vulnerability to agenda-manipulation in this sense. We have, however, already touched upon a theoretical concept the negation of which seems to capture the essence of this form of manipulation, viz. path-independence. If a procedure is path-independent, then obviously the winner cannot be affected by fiddling with the order in which the alternatives are introduced to the voting body. What path-independence specifically says, it will be recalled, is that if the alternative set is partitioned in various ways and the winners are picked from each subset to be considered with the rest of the alternatives, the winners in the set consisting of the subset winners and the rest of the alternatives are precisely the same as the winners if all the alternatives were considered simultaneously. Path-independence thus specifically rules out the possibility of affecting the outcomes by the partitioning of the alternative set in a suitable way. So we have to conclude that the possibilities for agenda-manipulation in this sense are almost ubiquitous as far as the procedures we have considered are concerned. The sole exception to this rule is approval voting, as we observed earlier.

SIXTH PROBLEM

But agenda-manipulation can take on other forms as well. Of these we shall consider the following one. It consists of introducing "straw-man" alternatives to defeat some other alternatives which in turn are wiped out by further alternatives in the original set, the idea being that the alternatives defeated by these artificial ones manage to beat those that would not be defeatable by the alternatives in the original set. The maneuvering thus involves an artificial enlargement of the alternative set so that the final winners will, however, be in the original set. In concrete decision making contexts this manipulation would involve e.g. introducing unrealistic alternatives in the voting sequence in order to exclude some realistic ones. The unrealistic alternatives, on the other hand, would eventually be defeated by some realistic ones.

Assuming that voting is sincere this second form of manipulation can be considered in the light of another theoretical property we have already considered, viz. WARP. Procedures satisfying WARP cannot possibly be manipulated in the above fashion as the property of heritage guarantees that every winner in the enlarged set must be a winner in all subsets it belongs to. Moreover, if in the "realistic" subset there is a tie between some alternatives, then all or none of them wins in the enlarged set. As we observed above, WARP is a rare property: of our procedures only approval voting satisfies it.

These two forms of agenda-manipulation probably do not exhaust the forms that can be encountered when observing the behaviour of the agenda-controllers in the real world voting bodies. But certainly many instances of agenda-manipulation take one of these two forms. While in the case of preference misrepresentation the problem can be avoided by restricting the information available of other voter's preferences, the undesirable features pertaining to agenda-manipulation can be practically eliminated by restricting the powers of the agenda-builder. For instance, the various speaker's rules applied in contemporary parliaments restrict the possibilities of speakers freely choosing the sequence of motions considered by the parliament. In the Finnish parliament, for instance, the rule is that the alternatives most widely apart should first be subjected to pairwise comparison in the application of the amendment procedure. To the extent that an unambiguous meaning can be given to the distance between alternatives, this rule completely eliminates the power of the speakers to manipulate the agenda.

It is also worth observing that some constitutional devices have been invented to overcome the problems of preference misrepresentation. In the presidential elections in France the method used is the plurality runoff system. As the saying goes, the voters are expected to vote "with their hearts" on the first round, but "with their brains" on the second. However, the proliferation of all kinds of opinion polls has made this idea of questionable validity. Rather, it has become ever more likely that large segments of the population vote "with their brains" on both rounds. Be that as it may, the French system prohibits the publication of opinion poll results within a restricted period before the elections. The idea behind this ruling is probably to diminish the likelihood of preference misrepresentation. So the problems related to these two forms of strategic behaviour have been known to the decision makers.

Let us now turn to a strategic property that is perhaps less known to real life decision making bodies but which can nevertheless be considered as a theoretical property of a similar genre as the misrepresentation of preferences and agenda-manipulation.

9.5. Sincere truncation of preferences

The sincere truncation of preferences pertains typically to procedures which utilize the entire preference orderings of the voters as input data. But as we have assumed in the preceding that the preference orderings are "given", we can generalize the discussion on preference truncation to all voting procedures so far considered. We shall then in effect assume that the voters give their preference orderings - possibly truncated - and the winner is then determined according to the principles of each procedure.

The notion of the sincere truncation of preferences was introduced by Steven J. Brams (1982) in connection with the evaluation of a nomination procedure used by a professional association. The procedure is that of Hare. What Brams specifically points out in his article is that the idea which seems at least partly to have been given as a justification of the procedure, viz. that it is never in a voter's interest not to indicate his entire preference ordering, simply does not stand critical evaluation. By way of counterexamples Brams shows that the nomination procedure used, i.e. Hare's method, is vulnerable to what he calls the sincere truncation of preferences. This means that under the procedure it may happen that a voter is better off if he does not indicate his entire preference ordering than if he does. One of Brams' counterexamples is the following.

Examples 9.12. $|N| = 21$, $X = \{a,b,c,d\}$

7 voters	6 voters	5 voters	3 voters
a	b	c	d
b	a	b	c
c	c	a	b
d	d	d	a

If everyone reveals his entire preference order, d will first be eliminated and three votes are transferred to c. Next b will be eliminated and its 6 votes transferred to a whereby a wins.

But suppose that the three voters ranking d first do not reveal their preference order any further, ceteris paribus. Alternative d is again eliminated first, but there is no candidate to whom the three votes could be transferred. Hence the next alternative to be eliminated is c and its 5 votes are transferred to b. Now b wins. As b is ranked higher than a by the three voters who rank d first, the truncation of preferences is obviously beneficial to them.

The notion of sincere preference truncation comes obviously very close to preference misrepresentation, but there is a definite difference: the truncation of preferences does not mean that voter i indicates that x P_i y when in fact his preferences is y P_i x. What the preference truncation means is that the voter sincerely indicates that y P_i x if both y and x are among the alternatives he ranks highest, but he does not report any preference between x and y if they are not among the alternatives ranked highest by him. In the latter case he actually lets the preferences of the others decide. So there is a difference between sincere preference truncation and preference misrepresentation. Actually the word "sincere" brings out this difference most clearly: there is nothing "insincere" in sincere preference truncation, while there is certainly something "insincere" in preference misrepresentation.

On closer scrutiny the difference between these two notions becomes somewhat smaller. What is at issue in sincere preference truncation is actually a restriction on preference misrepresentation: in determining which systems are vulnerable to sincere preference truncation one in effect asks in which systems there are situations such that one can benefit from indicating the indifference towards, instead of a strict preference between, any two alternatives. Obviously we are dealing with a specific type of "allowed" preference misrepresentation. Consequently, it is not at all clear if the systems that we have found vulnerable to preference misrepresentation sensu largo are also vulnerable to sincere preference truncation.

What then would be the practical import of this strategic concept? Obviously its main relevance is in systems operating on entire preference orderings or, more specifically, in procedures whose main motivation is that the voters' preference ranks are "duly" taken into account in the social choices made by the procedure in question. If it turns out that it could sometimes be beneficial for a voter to reveal only his first preferences, then this motivation would seem to lose some of its force. But the practical importance can be viewed from another angle as well. Obviously the voters are in general less concerned about the ordering between the alternatives that they rank low than about the order of highest ranked ones. It would follow then that the incentive for sincere preference truncation is there even without the possibility of benefiting from it. Thus, the fact that a procedure is vulnerable to sincere preference truncation adds to this incentive and works precisely in the opposite direction as the attempts to make the voters reveal their entire preference orderings.

To evaluate the vulnerability of the voting procedures to sincere preference truncation, we do not have many results which are directly applicable. As a matter of fact there is only one set of results on vulnerability to sincere preference truncation. The theorems are due to Fishburn and Brams (1984). To state them we need some definitions.

First we will define the moderate Condorcet condition MCC. A procedure F satisfies MCC iff for all alternative sets X and for all preference profiles R the following holds: $F(X,R) \subseteq \{x \in X |$ y $P_M x$ for no y in X$\}$ whenever the latter set is nonempty. Here P_M means that more voters strictly prefer the element indicated on the left side of the symbol to the element on the right side than vice versa. P_M is thus the symbol of plurality preference. The elements satisfying MCC are thus plurality undominated ones. A special case of MCC is the strong Condorcet condition SCC which we have already dealt with in section 5.1. As will be recalled the procedures satisfying SCC choose always precisely those alternatives that form the core of the voting game.

Now the first result of Fishburn and Brams states the following. Suppose that $|N| \geq 4$ and the alternative set consists of three elements only x_1, x_2 and x_3. Suppose, moreover, that the procedure we are considering is not vulnerable to sincere preference truncation. Under these conditions it cannot be the case that SCC is satisfied by the procedure.

The second theorem of Fishburn and Brams applies to situations which also contain three alternatives and in which the number of voters is at least seven. Under these conditions, the

SIXTH PROBLEM 131

theorem says, it cannot be the case that a procedure is not vulnerable to sincere preference truncation and satisfies MCC.

A glance at Table 5.2. in the preceding shows how helpful these results are in determining which of the procedures we have considered are vulnerable to sincere preference truncation. We observe that three procedures satisfy SCC and hence by the above theorem these procedures are vulnerable to sincere preference truncation under the stated conditions. In other words, if the number of voters is at least four and the number of alternatives is three, the maximin method along with Nanson's and Schwartz' procedures is vulnerable to sincere preference truncation. Examples of the vulnerability of these procedures and the amendment procedure, which also turns out to be vulnerable to sincere preference truncation, have been discussed by the present author elsewhere (Nurmi, 1984a; 1984b). The following example shows that Copeland's procedure which does not satisfy SCC is vulnerable to sincere preference truncation (Nurmi, 1984a).

Example 9.13. $|N| = 6$, $X = \{a,b,c,d\}$

person 1	persons 2 and 3	person 4	persons 5 and 6
a	c	b	b
d	a	d	c
b	d	a	d
c	b	c	a

Now the Copeland scores are: -1 for a, 1 for b, 1 for c and -1 for d. The result is then a tie between b and c. Person 1 can, however, improve upon this outcome by truncating his preference so as to state his first preference, a, only. Thereby the new scores are: -1 for a, 2 for b, 1 for c and -2 for d. The new outcome is thus b, clearly a preferable outcome for person 1 to the previous one (b-c tie). We are here making the rather natural assumption that person 1's truncated preference is honoured by the procedure so that e.g. in the comparison between a and b person 1 is interpreted as voting for a. Similarly he is interpreted as preferring a to all other alternatives. The only difference the truncation makes is that the voter in question is assumed to abstain in those pairwise comparisons which involve alternatives between which he has not indicated his preference.

Of the methods that satisfy the Condorcet winning criterion but not the strong Condorcet condition we have in addition to the amendment and Copeland's procedures to consider Black's and Dodgson's methods. The following example will show that both of them are vulnerable to sincere preference truncation.

Example 9.14. |N| = 3, X = {a,b,c,d}

person 1	person 2	person 3
a	c	b
b	b	c
c	d	d
d	a	a

In this example b is the Condorcet winner and will therefore be chosen by Black's method. If person 2 now truncates his preferences so as to indicate that c is his most preferred alternative, but gives no further information about his preferences, then b is no longer the Condorcet winner. Hence, the Borda winners will be chosen by Black's procedure. To determine the Borda winners in the case of a preference truncation, we assume that the median Borda score is given to those alternatives over which the voter indicates no preference. This means that in our example person 2's preference truncation is interpreted so that he gives 1 point to a, b and d each (for a discussion of the plausibility of this interpretation, see Nurmi, 1984a). Then the result is a tie between b and c, clearly an outcome that person 2 prefers to the outcome b. Notice that if person 2's preference truncation is taken to mean that the gives 0 points to a, b and d, then the conclusion concerning Black's method's vulnerability to preference truncation can be made a fortiori as now c becomes the winner after person 2's preference truncation.

The above example also shows that Dodgson's method can be manipulated by sincere preference truncation. If every person indicates his entire preference order, the outcome is b, the Condorcet winner. After person 2's preference truncation in the fashion described above, b is no longer the Condorcet winner and it takes one preference change to make it one, viz. person 1 has to change his mind about the preference between a and b. But c also can be made the Condorcet winner after the truncation by just one preference change, viz. by person 1's (or person 3's) change of mind between the preference order of c and b. Hence, there is a tie between b and c, after the truncation. And this, as we just observed, is clearly preferred by person 2 to the victory of b.

Now the question arises as to how the procedures which do not satisfy the Condorcet winning criterion perform. First we observe that the question is not meaningful as far as plurality and plurality runoff methods are concerned as when these procedures are utilized the voter in fact only reveals his first preference. In other words, these procedures, trivially, are not vulnerable to sincere preference truncation.

Also the Borda count - which as was pointed out above does not satisfy the Condorcet winning criterion - is vulnerable to sincere preference truncation. This can be seen from Example 9.14. where b is the Borda winner. If now person 2 truncates his preferences so as to indicate no more than that c is his most-preferred alternative, the result is a tie between b and c if the procedure mentioned above is resorted to in assigning the Borda scores to alternatives over which the voter does not indicate the preference order. Clearly, this outcome is better from person 2's view-point than the victory of b. Hence the vulnerability of the Borda count to sincere preference truncation follows.

This leaves us two procedures to consider, viz. Hare's and Coombs'. As we already noticed Brams' (1982) discussion has shown that the former is vulnerable to sincere preference truncation. About the latter I venture the conjecture that it is not vulnerable to sincere preference truncation (see Nurmi, 1984b). Although no proof to that effect will be presented, the conclusion seems plausible enough in view of the modus operandi of Coombs' procedure: both the upper and lower end of each individual's preference order is important as the winning criterion is determined on the basis of the distribution of first preferences, while the elimination of alternatives is determined on the basis of the distribution of last preferences. Thus, by truncating his preferences the voter actually loses the possibility of determining the elimination of his worst alternative. Hence, the difference he would seem to be able to make is the worse for himself.

It would seem, then, that plurality, plurality runoff and Coombs' procedure are in their own class as far as vulnerability to sincere preference truncation is concerned. All other methods can be manipulated by the sincere truncation of preferences. But how important is this property in practice? Not necessarily very important as the possibility of benefiting from sincere preference truncation requires fairly detailed information about the preferences of the others. One should notice that sincere preference truncation is a less general way of benefiting from giving distorted information than the misrepresentation of individual preferences because every method that is vulnerable to the former kind of strategic behaviour is also vulnerable to the latter, while the converse is not true. In each situation where an individual can benefit from truncating his preferences, he can also benefit from misrepresenting them. So the remarks presented above concerning the difficulty of benefiting from preference misrepresentation apply to the manipulation via sincere preference truncation as well. More specifically the hierarchy of preference misrepresentation holds for the sincere preference truncation as well.

Thus, even though all the procedures we have considered are in principle vulnerable to preference misrepresentation and most of them to agenda-manipulation and sincere preference truncation, there are certainly differences between them concerning the difficulty of benefiting from these types of strategic devices. Moreover, various practical rulings have been instituted to eliminate or at least considerably diminish the possibilities of advantageous strategic behaviour. Although the sincere preference truncation has not yet played a very great role in practice, one could easily envision devices to eliminate incentives to truncate one's preferences. A very simple one, of course, would be to exclude all such ballots which do not indicate the entire preference schedule of the voter. This kind of ruling would, of course, be rather radical and perhaps unacceptable on other grounds. A gentler way of doing away with incentives to benefit from sincere preference truncation would be to order the truncated places in a voter's preference order in a random fashion. This would perhaps be more in accordance with the most common motivation underlying the preference truncation: viz. that the person really does not care in which order the alternatives in the lower end of his schedule are placed.

CHAPTER 10

SOCIAL CHOICE METHODS BASED ON MORE DETAILED INFORMATION ABOUT INDIVIDUAL PREFERENCES

All the results pertaining to the performance of the voting procedures with respect to various criteria discussed above take their point of departure from the assumption that individuals have a preference relation over the set of alternatives. There certainly is considerable plausibility in this assumption for a wide class of choice situations. But often one could well ask whether this assumption contains all the relevant information about individual tastes, wishes etc. that would seem to be called for.

Sometimes the mere preference order of individuals does not seem enough as a basis for determining social choices. For instance, various preference comparison considerations might make it plausible to ask people to give more detailed information about their preferences than the order of priority in which they would put the alternatives. If in a two-person society person A very mildly prefers x to y in a situation where only these alternatives are available, while for person B the choice of y over x is a matter of life and death, then there would seem to be a case for arguing that y should be preferred to x by the society in toto. This case, in turn, is built on other than merely preference order considerations.

But in other cases, asking individuals' preference orders would seem to be too much. Often people are not really interested in pondering upon which of two quite intolerable alternatives is less intolerable. In such cases the assumption of an individual preference relation as a complete and transitive one is too stringent. In the following we shall take a look at social choice methods that take into consideration more than the mere preference order over alternatives. In the next chapter we shall turn to a consideration of some recent results pertaining to social choice when the individuals are assumed to have somewhat less definite views than preference orders over the alternatives.

There are several ways of resolving the social choice problem of the above two-person society in a manner that results in the choice of y which would intuitively seem to be the "just" outcome: either resort to a non-anonymous procedure which would give more weight to the person's preference whose intensity of preference is stronger or resort to a non-neutral one which gives more weight to an alternative which is more intensely preferred. Both of these ways require a yardstick of preference intensity that would allow an interpersonal comparison of the preference intensity. Most common among such yardsticks is the von Neumann-Morgenstern utility, that is, a function defined over the alternatives such that certain requirements are fulfilled (or certain axioms hold).

10.1. The von Neumann-Morgenstern utility and classes of interpersonal comparability

The von Neumann-Morgenstern utility concept was introduced in the classic book (von Neumann and Morgenstern, 1944) more than forty years ago. In the following we shall, however, mainly follow the presentation of Harsanyi (1977). Consider a risky prospect

$$C = (A,p; B,1-p)$$

i.e. a lottery where the probability of prize A is p and the probability of prize B is 1-p ($0 \leq p \leq 1$). We say that a function u defined for any risky prospect has the expected utility property if it assigns to each risky prospect the utility that equals the expected value of the prizes in the corresponding lottery. In other words the utility function u has the expected utility property if for all prizes A, B and C and for all probability values p, the following holds:

$$u(C) = pu(A) + (1-p)u(B)$$

whenever $C = (A,p; B,1-p)$.

Now if a utility function has the expected utility property, then it is called the von Neumann-Morgenstern utility. If we define the preference of D_1 over D_2 so that the person having the preference in question always (i.e. with probability one) chooses D_1 when faced with the possibility of choosing either D_1 or D_2, and extend this notion of preference to risky prospects as well, then preference based behaviour can be described *as if* the person in question were maximizing his utility function over the alternatives at hand.

USING MORE THAN PREFERENCE ORDERS

What then are the conditions under which a utility function with the expected utility property exists? In other words, what are the conditions that guarantee that when faced with the choice between risky prospects the person chooses as if he were maximizing his expected utility function over the prospects? The answer to this question gives the conditions under which a person's choice behaviour can be represented by a utility function with the expected utility property.

The following list of axioms will guarantee the existence of such a function:

1. Every certain outcome A can be described as a risky prospect:

$$A = (A,1; B,0).$$

2. For all A, B and p:

$$(A,p; B,1-p) = (B,1-p; A,p).$$

3. Let $A_1 = A_2$. Then

$$(A_1,p_1; A_2,p_2; A_3,p_3; \ldots ; A_k,p_k)$$
$$= (A_1,p_1+p_2; A_3,p_3; \ldots ; A_k,p_k).$$

4. Let Z be the set of all risky prospects. Then person i has a preference relation R_i over Z such that R_i is complete and transitive.

5. Let $I_{BAC} = \{p \mid B\ R_i\ (A,p; C,1-p)\}$

and $S_{BAC} = \{q \mid (A,q; C,1-q)\ R_i\ B\}$

i.e. I_{BAC} (S_{BAC}, respectively) consists of those probability values p (q) for which the outcome B (prospect (A,q; C,1-q)) is preferred to prospect (A,p; C,1-p) (outcome B). The axiom requires that for all A, B and C, the sets I_{BAC} and S_{BAC} are closed sets. Intuitively this means that a small change in the value of p is accompanied by a small change in the utility value of the risky prospect.

6. Let $A\ R_i\ B$ and $p > 0$. Then

(A,p; C,1-p) R_i (B,p; C,1-p) and conversely, i.e. if (A,p; C,1-p) R_i (B,p; C,1-p), then A R_i B.

The axioms 1.-6. imply the existence of a utility function having the expected utility property such that the function represents the individual's preferences.

Now obviously the fact that a utility function exists for all individuals does not imply that the utility values are in some sense comparable. Indeed, the satisfaction of the above axioms does not guarantee a unique utility scale for any individual. The von Neumann-Morgenstern utilities are unique up to an affine transformation. This means for example that if u is a utility function over a set X of alternatives then also v is a utility function if

$$v(x) = k\,u(x) + b,$$

for all x in X where k and b are constants and k > 0.

As a step towards the interpersonal comparability of utilities one could resort to normalization where the end points of the individual utility scales are fixed e.g. so that the worst alternative is given the value 0 and the best the value 1. This would be the 0-1 normalization. Riker and Ordeshook (1973, p. 38) suggest a way of finding the utility values of alternatives for a finite alternative set X. Axiom 4. above guarantees that the set of highest ranked alternatives and the lowest ranked ones are both nonempty for all individuals. Let us give these alternatives the values 1 and 0, respectively. For any alternative x_j between these extreme ranks we offer the individual the choice between the following two alternatives:

1) x_j with certainty, and

2) the lottery (A,p; B,1-p) where A is the highest ranked alternative (or one of them in case there are many) and B is the lowest ranked one (or one of them).

Axioms 1. and 5. guarantee that there is a value of p such that the individual is indifferent in choosing between options 1) and 2). This, in turn, means that he assigns to each of these options the same utility. But we already know the utility of 2) because the utility function has the expected utility property by virtue of satisfying 1.-6. This utility is, of course,

$$p\,u(A) + (1-p)\,u(B).$$

In the special case of 0-1 normalization u(A)=1 and u(B)=0 and hence $u(x_j)$=p.

But this step towards interpersonal comparability does not take us far enough. It may still be the case that both the utility values and the utility differences differ from one individual to another. Suppose that one alternative in X implies the death of one person, say i, in the group making the choice between alternatives, while the other alternatives make only a minor change in the welfare levels of the individuals. Then obviously both the level and the utility differences between alternatives would be different for i and the other members of the group in the sense that i would probably assign a much larger difference between the worst and best alternative than the other individuals. Similarly, i would regard the utility of his worst outcome as much smaller than any other person in the group would regard the worst outcome even though by von Neumann-Morgenstern normalization all would give the same numerical value to it.

More generally interpersonal comparability can be dealt with by considering the transformations in individual utility functions that leave the social preference relation unaffected (see Roberts, 1980). In other words, we shall focus on transformations of the following type: $u'_i(x) = f_i(u_i(x))$ which do not affect the social preference in the sense that the social preference constructed from or determined by u'_i and u_i (i=1, ..., n) remains the same, i.e. R' = R over the set X of alternatives.

The case of ordinality and interpersonal non-comparability is characterized by invariance transformations that are independent, and strictly monotonically increasing. This is the widest class of transformations considered here. Independence means that f_i can be different for each individual, while strict monotonicity means that for all i in N and for all x and y in X: whenever $u_i(x) > u_i(y)$ then $u'_i(x) > u'_i(y)$, u'_i being determined by $u'_i(x) = f_i(u_i(x))$.

The von Neumann-Morgenstern utility is a restriction to this class of invariance transformations, viz. the transformation $f_i(u_i(x)) = u'_i(x) = k_i u_i(x) + b_i$, with k_i and b_i constants and $k_i > 0$, is the allowed transformation that leaves the social preference unaffected. In this case we have cardinality but no interpersonal comparability.

Now the individual welfare levels may be comparable without there being an underlying cardinal utility function. For example, the social preference of social states could be determined by comparing the welfare levels of the worst-off individuals in these states, somewhat like Rawls (1971) has suggested. Ordinality and interpersonal level-comparability obtain when the allowed transformations f_i are identical for all individuals and moreover strictly monotonic.

The next class of utility functions consists of ones that leave the social preference unaffected if and only if the transformations f_i are independent of i and strictly positive affine, that is, $u'_i(x) = k\, u_i(x) + b_i$, where k does not depend on i. In this case the utility function allows for the interpersonal comparison of welfare gains. Thus we have both cardinality and interpersonal comparability. Both are required if utilitarian social choice methods are to be used. In other words, it makes no sense to use utilitarian methods unless the individual utilities satisfy both cardinality and interpersonal comparability.

Having thus briefly looked into various types of individual utility functions, we are now in a position to see how they can be applied in making social choices, i.e. how one can utilize the additional information provided by the functions.

10.2. Old and new methods

The most straight-forward use of the information provided by individual utility functions over a set of alternatives is made by utilitarians who in the spirit of Bentham regard the alternative with the largest sum of individual utilities as socially best (Riker, 1982). Denote by $u_i(x)$ the utility value assigned by individual i to alternative x. Then $\Sigma_i\, u_i(x)$ is the sum of the utilities given to x by all individuals. The Bentham method is now the social choice function F such that

$$F(X, U) = \{x_j \in X | \Sigma_i\, u_i(x_j) = \bar{u}\}$$

where U is the n-tuple of individual utility functions over X, the set of alternatives, and $\bar{u} = \max_j \Sigma_i\, u_i(x_j)$.

Now in utilitarian theory one does not have to assume that the utilities considered in the above expressions are normalized by fixing the maximum and minimum values as explained in the previous section. In other words, in utilitarian thinking one could well deal with utility scales that are unique up to identity transformation. This would mean that we are using absolute utility values. If that would be the case, then there obviously could well be individuals whose maximum utility value for an alternative in X would be much higher than the maximum value of some other person. But obviously there would be difficult problems in ascertaining these absolute utility values. Hence we assume that the utilities above denote the von Neumann-Morgenstern utilities.

Another rather similar method discussed by Riker (1982,

p. 31) is the Nash method which determines the winner as follows (see also Dyer and Miles, 1976):

$$F(X,U) = \{x_j \in X | \pi_i \, u_i(x_j) = u'\}$$

where $u' = \max_j \pi_i \, u_i(x_j)$. Thus, the Nash method chooses those alternatives for which the product of individual utilities is at the maximum. Obviously, this method is very sensitive to zero values of utility: in 0-1 normalization no such alternative could be chosen before the others that would be ranked worst by at least one individual. To avoid this intuitively unreasonable feature, one could resort to 0.5 - 1 normalization.

Both the Bentham and Nash methods utilize the utility values given by individuals to alternatives. In a sense the Borda count does the same: it assigns each individual's highest ranked alternative the utility value of k-1 if the number of alternatives is k, the next highest alternative the value k-2 and so on. One could argue that the Borda count utilizes the k-1 - 0 normalized utility values. This, however, is not the case as the Borda count does not take into account the intensity of preferences at all. It assumes that differences between utility values are the same throughout the preference ordering. In other words, it is assumed that the worst alternative is given the value of zero, the next alternative the value one, the next the value of one plus the value of the preceding alternative and so on. This obviously is a special case of individual utility assignment and cannot be assumed to be valid in general. Hence, the Borda scores cannot be interpreted as utility values.

The method of reaching consensus in group decision making presented by Lehrer and Wagner (1981) introduces another additional consideration into group decision making, viz. the opinion of voters concerning each other's expertise and/or judgement in the issue at hand. The most natural context in which the Lehrer-Wagner method could be utilized is one in which a group of experts representing somewhat different fields tries to reach a consensus e.g. about the action the group should take. Examples of this kind of decision making situation include the choice of therapy for a patient, the allocation of resources to large inter-disciplinary research projects and the like.

A typical feature of these settings is that the decision making group consists of competent representatives of various disciplines or fields of interest. The idea underlying the method is that these experts have opinions both about each other's expertise and about the policy or action alternatives. Both of these considerations play a role in the process of reaching what Lehrer and Wagner call rational consensus.

The method can be applied with or without information about the individual utilities of alternatives. The basic notion in the Lehrer-Wagner method is the weight matrix which is a n-by-n matrix W consisting of entries w_{ij} indicating the weight assigned by individual i to individual j (i, j = 1, ..., n). In other words, w_{ij} indicates the opinion held by i concerning the reliability, expertise or some similar property of j in making the judgement in the issue to be decided. The constraints on w_{ij} are the following:

(1) $0 \leq w_{ij} \leq 1$, and

(2) $\Sigma_j w_{ij} = 1$, for all i and j.

The matrix W thus indicates the weights given by voters to each other. The matrix product $W \times W = W^2$, in turn, is one consisting of entries that are weighted means of the rows (or columns) of W. The weights are given by the corresponding columns (or rows).

Now raising W to higher powers we may eventually come up with a matrix W^s with identical rows. Lehrer and Wagner interpret this convergence matrix as a consensus one where each row indicates the consensual weights, i.e. each row is a vector of consensual weights reflecting the group's opinion of the expertise of the members. The conditions under which W^t converges as t increases are not very stringent and we shall therefore assume that it always converges unless specifically otherwise stated.

There are two variations of this what Lehrer and Wagner call elementary model of reaching consensus. The first one could be called the model of indirect democracy where the group members choose from among themselves a representative to make choices or decisions on behalf of the group in toto. The other variation resorts to utilities as given by the voters regarding the alternatives under consideration. In both variations, however, the weight matrix and more specifically the vector of consensual weights - i.e. a row of the convergence matrix - play a crucial role.

In the former variation one simply chooses the voter whose weight in the consensual weight vector is the largest. In the latter variation, we consider the utility matrix

$$A = \begin{bmatrix} a_{11} & \cdots & a_{1k} \\ a_{21} & \cdots & a_{2k} \\ \vdots & & \vdots \\ a_{n1} & \cdots & a_{nk} \end{bmatrix}$$

where a_{ij} (i = 1, ..., n; j = 1, ..., k) denotes the utility of alternative j to voter i. Let the consensual weight vector be $C = [w_1, ..., w_n]$. The vector CA denotes the consensual utilities of alternatives. An obvious candidate for choice would, then, be the alternative with the largest consensual utility.

Let us finally consider the leximin rule (see e.g. Kim and Roush 1980, p. 69) which does not require that the utilities have more than ordinal significance, i.e. the utility functions have to satisfy ordinality with interpersonal comparability only. For all alternatives x one forms a vector of individual utilities

$$u(x) = [u_1(x), u_2(x), ..., u_n(x)],$$

where the utilities are arranged in nondecreasing order, i.e. $u_1(x)$ is less than or equal to $u_2(x)$, the latter, in turn, less than or equal to $u_3(x)$ etc. The choice set is determined on the basis of a comparison of the vectors u(x) thus formed.

Consider first two alternatives x and y in X. x is then lexicographically preferred to y iff $u_i(x) > u_i(y)$ for some i in the vectors and $u_j(x) = u_j(y)$ for all j = 1, ..., i-1. Obviously if i=1 it means that the least utility value of x is strictly larger than the least utility value for y. Obviously, the set of maximal elements with respect to this lexicographic preference is always nonempty. This set is then the choice set under the leximin rule.

10.3. An assessment

An obvious way to evaluate the methods outlined in the preceding section would be to use the criteria employed earlier, all of which could be applied. We shall, however, refrain from that kind of assessment and concentrate on problems that are more specific to the methods at hand (see Nurmi, 1985). The most crucial of these pertains to finding out the individual utilities. The problem is: how to determine the individual utilities in a fashion that would not undermine the very rationale of using utility-based preference aggregation?

Of course, it could be protested that the methods discussed in this chapter are not intended for contexts involving interest or opinion aggregation but for those where the aim is to find out what is "best" for the collectivity. In the latter types of situations - it could be argued - the approach is that of a benevolent dictator and the procedures reflect the various ways in which the "best" could be interpreted. To this line of reasoning one can present counterarguments. Firstly, if the outcome resulting e.g. from the application of the leximin rule is "best" by definition, then of course the case is closed and no discussion about the virtues and shortcomings of the rule is possible. This is hardly a reasonable standpoint given the fact that there are many rules proposed for similar situations. Secondly, the violation of e.g. the Condorcet loser criterion is a serious shortcoming even in cases where the voters are interpreted as criteria and the problem is to find out the best alternatives in terms of the criteria. The same applies to the other considerations we have used in the evaluation of the procedures with the obvious exception of manipulability which has no counterpart in the multiple criterion choice problem.

If we first consider the Bentham and Nash methods, it seems that they do not work if the von Neumann-Morgenstern utilities are used. The reason is simply that even if the same normalization is used for all individuals, the utilities are not interpersonally comparable, as was pointed out in the preceding section. Consequently, the methods that choose the alternatives having the largest sum or product scores in terms of the von Neumann-Morgenstern utilities do not necessarily maximize the "absolute" sum or product of utilities.

But suppose we have the option of choosing between dropping the cardinality requirement underlying the von Neumann-Morgenstern utilities and being satisfied with ordinal scales only, or making more stringent uniqueness requirements on the utility scales. Obviously the former alternative is feasible for one procedure only, viz. the leximin rule, as in all others we must perform mathematical operations (addition and multiplication) that do not make sense in ordinal scale variables. However, thereby we encounter new problems. Consider the following example.

Example 10.1. $|N| = 3$, $X = \{x_1, x_2, x_3, x_4\}$

person 1	person 2	person 3
x_1	x_1	x_4
x_2	x_4	x_2
x_3	x_3	x_3
x_4	x_2	x_1

As the order of alternatives and that only is significant, we can give the last ranked alternative an ordinal utility value of, say, 1, the next ranked alternative the value 2, etc. The leximin utility vectors of alternatives then become:

USING MORE THAN PREFERENCE ORDERS

$$u(x_1) = (1,4,4)$$
$$u(x_2) = (1,3,3)$$
$$u(x_3) = (2,2,2)$$
$$u(x_4) = (1,3,4)$$

x_3 being the winner. Consider now the subset $A = \{x_1, x_3\}$. Obviously in this subset:

$$u(x_1) = (1,2,2) \text{ and }$$
$$u(x_3) = (1,1,2)$$

making x_1 the winner. Hence the independence of irrelevant alternatives is violated.

The leximin rule could, thus, be considered as one of the "ordinal" methods we have considered in the preceding. But with equal justification it can be considered as one of the methods asking the voters to indicate more than just their ordinal preference orders concerning the alternatives. We observe that the performance of the leximin rule would be reasonably good if it were assessed in terms of the criteria we have been using in the context of the previous chapters. Thus, the leximin rule (its ordinal version) satisfies the Pareto criteria and monotonicity. Furthermore it is consistent, but fails on WARP and on path-independence.

Its weak points are the Condorcet-criteria and the majority criterion. Example 10.1. illustrates some of its properties. In that example there is a Condorcet-winner, viz. x_1. It is, however, not chosen by the leximin rule. Indeed, the rule chooses the Condorcet loser in the example, viz. x_3.

The same example can be resorted to in order to illustrate the violation of heritage which implies the violation of both WARP and path-independence. Consider the subset $\{x_1, x_3, x_4\}$ of X. For this subset the leximin utility vectors are:

$$u(x_1) = (1,3,3)$$
$$u(x_3) = (1,2,2)$$
$$u(x_4) = (1,2,3)$$

Hence, x_1 is chosen thus violating the heritage condition, which requires that x_3, the over-all winner, should be the winner in every subset of alternatives to which it belongs.

The failure of the leximin rule to satisfy the majority winning criterion can also be seen from Example 10.1. There x_1 is the majority winning alternative and yet x_3 is chosen by the rule. Indeed, this feature could be dramatized as much as one wished by multiplying the number of persons having the same preference profile as person 1 (or person 2) by 1000 or whatever number. Yet the choice would remain x_3 as long as no other changes were made.

Now what about the question of requiring more stringent uniqueness from the utility scales? Several problems are encountered when proceeding in this direction. The first is the fact that we do not have a well-established methodology for estimating the more unique than von Neumann-Morgenstern type utilities in practice. This problem is aggravated by the obvious vistas that are thereby opened for preference misrepresentation. Indeed, when the Bentham or Nash methods are utilized it would never hurt a voter to underestimate the utility he gives to his lowest ranked alternative. Nor would it do any harm for him to overestimate the utility of his highest ranked alternatives. In fact, one could go as far as to claim that the strategy of over- and underestimating one's preferences in this fashion dominates the strategy of sincere utility-revelation when these two methods are used.

But obviously this applies to the Lehrer-Wagner method as well. It would thus seem that to use these methods with more unique than von Neumann-Morgenstern utilities is a direct invitation to strategic preference misrepresentation. What makes this feature particularly serious is that no information about the utilities of others is needed for this kind of manipulation.

The Nash method has the further drawback that it is particularly vulnerable to the assignment of zero utilities: by assigning a zero utility to an alternative, any voter can render its total utility score zero, obviously a somewhat dubious property. Similarly, the use of negative values gives a voter disproportionate power in determining the outcome.

The Lehrer-Wagner method, in turn, has an additional possibility for strategic manipulation, viz. the weight assignment. It would seem never to be harmful for a voter to give himself more weight than the sincere weight assignment would imply, provided that the latter yields a value less than unity. Similarly, if one knows the persons who have diametrically opposed interests to one's own, then assigning a zero weight to these persons would never hurt a voter even though one's opinion of their expertise would imply a sincere weight assignment of something else than zero.

In sum, the most serious drawback of methods utilizing information other than that contained in individual preference orderings, is the fact that entirely new possibilities for the strategic manipulation of individual inputs are opened. As we saw above, the von Neumann-Morgenstern utilities which are by far the best-known utility functions will not work in the social welfare functionals we have considered. On the other hand, very little is known about the possibilities of eliciting a truthful utility-revelation in contexts involving genuinely political decision making. Let us, therefore, focus out attention on methods that ask the voters to give less than their preference orderings.

CHAPTER 11

ASKING FOR LESS THAN INDIVIDUAL PREFERENCE ORDERINGS

Now the least one asks of a voter is to indicate one alternative, either the one he deems best or the one he considers the worst. The former type of input is used in the plurality method, whereas the so-called negative voting utilizes the latter type of information. However, in negative voting the only sensible winning criterion would be the least number of negative votes. This, in turn, is bound to lead either to a large choice set or to several stages in a similar fashion as in Coombs' procedure.

But obviously one could ask the voters to reveal a fixed number of their highest ranked alternatives, e.g. to indicate the three alternatives they consider best either with or without indicating their preference order among them. This procedure is particularly popular in voting bodies which have been given the task of nominating three office holders or - perhaps even more commonly - of indicating a collective preference order among the three alternatives considered as best by the body.

An example of the latter variant is used in submitting to the nominating authority a proposal of a lower level body concerning the order of priority among the applicants to the office in question. The nominating officer may then have the option among the three candidates thus named or alternatively be entirely free to nominate anyone of the applicants. Obviously, in the latter case the importance of the priority order of the lower body may not be very great. But in either case it is of course important to determine whether there are non-arbitrary ways of coming up with a collective preference order concerning a subset of alternatives when the cardinality of the subset is fixed by the information asked of the voters.

USING LESS THAN PREFERENCE ORDERS 149

11.1. Constructing a social preference order for a subset of alternatives

Among the methods we have considered in the preceding there are some which do not naturally lend themselves to constructing a social preference order. For example, the amendment procedure is such a one; it produces a social winner, but does not automatically give any ordering between the rest of the alternatives. Similarly, Schwartz' procedure is primarily a method for reaching a social choice set.

But also among the methods that by their nature are capable of producing a social preference ordering, there are some that behave in a different manner when the voters are asked to give their entire preference orderings rather than to indicate their three or some other fixed number of highest ranked alternatives. If the voters are simply asked to list their three most-preferred alternatives without giving any order of priority between them, then the most natural way of choosing those three alternatives, over which the social preference relation is to be constructed, is to take the alternatives which have been mentioned by most voters in their lists of most preferred alternatives. Should this method be resorted to, then obviously if there is a majority alternative in the voters' underlying preference orderings, this will not necessarily be among the three alternatives that will subjected to further consideration. Let us consider the following example.

Example 11.1. $|N| = 5$, $X = \{a,b,c,d\}$

1 person	1 person	1 person	2 persons
a	a	a	b
d	d	b	c
b	c	c	d
c	b	d	a

Now if each person sincerely reports his three most-preferred alternatives, alternatives b, c and d will be chosen for further scrutiny and yet a has the majority of first ranks. This example shows that even though there is a majority alternative, it may not be chosen because the entire preference orderings of voters are not used. This possibility exists no matter which procedure is then utilized to obtain the social preference ordering over the three alternatives. This is simply due to the fact that a, the majority alternative, is not among the three alternatives that the voters mention most often. In other words, even if the method used in the last stage were such that it guarantees the choice of the majority candidate whenever one exists, the procedure as a whole may fail to accomplish this when the entire preference orderings of voters are taken into account.

A fortiori, the example shows that no procedure used for determining the social preference ordering over the set of three alternatives by asking persons to indicate their three most-preferred alternatives without their preferences over them, satisfies the Condorcet winning criterion when the entire preference orderings are considered. This example shows that in considering a fixed number of most-preferred alternatives only, one may lose the otherwise nice properties one's procedure might have. It should be observed that the problem here has nothing to do with utility-maximization on the group level. What is assumed is merely the individual preference orderings over the alternatives and nothing more.

The feature shown in Example 11.1. is certainly a drawback for procedures which would otherwise necessarily choose the majority alternative and Condorcet winner when one exists. It is, however, worth emphasizing that this result is based on two assumptions:

1) that the voters have an ordering over the entire set of alternatives, and

2) that they are asked to name a fixed number of alternatives without indicating their preference over this set.

In a sense the result is not all that surprising. What it says is that when the information about individual preferences is "positionalized" in an arbitrary fashion, the binary comparison criteria may give different results than when applied to non-reduced preferences.

What would happen then if, in addition to naming three alternatives, the voters were asked to reveal their preferences over the indicated alternatives? Obviously this situation would immediately lead to further questions concerning the procedure used in aggregating the truncated preferences thus obtained. But regardless of the procedure utilized in this further processing of the preferences, we may observe that it now becomes possible that a Condorcet winning alternative will not be among the set of alternatives named by the voters. That is, it may happen that no voter names the Condorcet winning alternative either as his first, second or third ranked alternative even though there is one. The following example illustrates this peculiarity.

USING LESS THAN PREFERENCE ORDERS 151

Example 11.2. |N| = 3, X = {a,b,c,d,e,f,g,h,i,j}

voter 1	voter 2	voter 3
e	g	h
d	f	i
a	j	c
b	b	b
c	c	a
f	a	e
g	e	d
h	d	j
i	h	f
j	i	g

Here b is the Condorcet winner, but all voters consider it worse than their third-ranked alternative. Hence, with sincere voting no voter includes b among his three best alternatives. So, no matter which procedure is used to come up with the social choice after the preference lists have been given by the voters, the Condorcet winner is not among the winners.

A glance at example 11.2. also shows that by restricting the preference revelation to the three highest ranked alternatives one does not guarantee the exclusion of the Condorcet loser when such an alternative exists. In the above example j is the Condorcet loser and yet it is named by voter 2. The following example makes this point more conspicuous.

Example 11.3. |N| = 5, X = {a,b,c,d,e}

voter 1	voter 2	voter 3	voters 4 and 5
a	b	c	d
b	c	a	c
c	a	b	b
e	e	e	e
d	d	d	a

Here d, the Condorcet loser, is ranked first by more voters than any other alternative. Obviously, the restriction does nothing to eliminate Condorcet losers.

The above remarks do not assume that any particular preference aggregation method will be used once the restricted preference orders have been given by the voters. But can one work out a collective preference order over, say, three or four collectively best alternatives? The following methods have been used in practice:

(1) Count for each named alternative the number of its first ranks. If there is an alternative with more than 50 % of the first ranks, rank this first in the social preference order. If no such alternative can be found, the following scoring function is used: each alternative is given 1 point for each first rank it gets, 1/2 points for each second rank it gets and 1/3 points for each third rank. The points are added up for each alternative and the alternative with the largest total score is given the first rank in the social preference order. One then determines whether among the remaining k-1 alternatives there is one that is ranked first by more than 1/3 of the voters. If there is one, it is given the second rank in the social ordering. Otherwise the second ranked alternative is determined by the scores of alternatives as in the case of the first rank. In all cases the third rank is determined by the scores.

The method is used e.g. in making nomination proposals for chancellorships in Finnish universities. More specifically, the consistoria or councils make their proposals concerning the person to be nominated as chancellor by indicating an order of preference among three candidates. The method described above is explicitly designed to guarantee "proportionality" in elections even though party or other kinds of lists are considered to be inappropriate. The nomination has to be made from the list of three candidates thus obtained, but the nomination decision-maker does not have to choose the first ranked candidate. In the particular case of university chancellors the nomination decision is made by the president of Finland. Thus although he is not completely constrained by the proposal produced by the collective body, the procedure is designed to indicate the preference of that body.

As was just said, this method is designed to guarantee some degree of proportionality in the sense that even if there is a large group of voters with identical preferences, it does not automatically dictate the social ordering apart from the first rank. We immediately observe that the method looks very much like the Borda count, but on closer inspection it is not equivalent to it. The present method is not reducible to any such scoring function that would give a points to the third ranked alternative, a+b to the second ranked one and a+2b to the first ranked one, as Borda required. Nonetheless, this method has similar peculiarities to the Borda count. The following example sheds some light on them.

USING LESS THAN PREFERENCE ORDERS

Example 11.4. $|N| = 10$, $X = \{a,b,c,d\}$

3 voters	3 voters	3 voters	1 voter
a	d	c	b
b	a	d	c
c	b	a	d
d	c	b	a

Suppose now that the above method were utilized so that for each first rank an alternative is given 1 point, for the second rank 1/2 points, for the third rank 1/3 points and for the fourth rank 1/4 points. The total scores of the alternatives would then be:

a: $5\frac{9}{12}$

b: $4\frac{1}{4}$

c: $5\frac{3}{12}$

d: $5\frac{7}{12}$

Suppose now that b were left out of the contest for some reason (or that the method were used to eliminate the alternative with the lowest total score) and that the method were again applied to the set of remaining alternatives to produce the social ordering over three alternatives. We would then have the following preference configuration:

3 voters	3 voters	3 voters	1 voters
a	d	c	c
c	a	d	d
d	c	a	a

That is, the preferences are identical except that b has been erased from each preference order. Using the above method we get the following scores:

a: $5\frac{5}{6}$

c: $6\frac{1}{2}$

d: 6

In other words, the social preference order adc has been reversed to cda by the elimination of b, the alternative with the lowest score in the 4-alternative contest. There certainly does not seem to be much point in saying that cda is the "right" social preference order. Similar behaviour is characteristic of the Borda count as well, as has been shown by Fishburn (1974).

(2) The second method starts by assigning the alternative with the largest number of first ranks in individual ballots to the first rank in the social preference order. Thereafter, one adds up the number of first and second ranks given to various alternatives and assigns the alternative with the largest sum score to the second rank in the social preference order. The third rank is given to the alternative whose sum score when counting the number of first, second and third ranks is largest.

This method is used in similar contexts as the previous one, that is, in making nomination proposals. This time the offices to be filled are those of a bishop in the Lutheran church of Finland. In each district the entitled voters can vote for three candidates indicating their order of preference. The nomination proposal is then computed from the votes in the above fashion.

This method is usually considered to be majoritarian in spirit, that is, no proportionality considerations are intended to enter these elections. In case there is an alternative that is ranked first by more than 50 % of the voters methods (1) and (2) rank it first, but in the absence of such an alternative marked differences may occur. Consider the following example.

Example 11.5. $|N| = 27$, $X = \{a,b,c\}$

11 voters	10 voters	6 voters
a	b	c
c	c	b
b	a	a

According to method (2) the social preference order is the following: a c b, because a has the largest number of first ranks, c the largest number of first and second ranks among b and c, and b then is ranked third.

On the other hand, method (1) would produce the following social preference order: b a c, because b has the largest score ($16\frac{2}{3}$), a has more than 1/3 of the first ranks and finally comes c.

If the number of persons with the preference order cba were 8 instead of 6, the social preference order determined by method (1) would become: c a b as now c would have the largest score ($18\frac{1}{7}$) and a would again have more than 1/3 of the first ranks which would leave b the third ranked alternative. We witness thus

USING LESS THAN PREFERENCE ORDERS   155

a complete reversal of the social preference even though not more than 2 persons have been added to the voting body.

The preceding discussion shows that if one wished to guarantee the satisfaction of Condorcet criteria or other related properties of social choices, then the restriction to a subset of alternatives does not seem a workable solution. If people are asked to reveal their top-most alternatives only - with or without the preference order among them - the information on paired comparisons is lost. This, in turn, can lead to a situation where a Condorcet winner is not among the subset indicated by the voters. But what if the voters do not have a preference order over the set of alternatives to start with? All our previous results rest on this assumption. It may, therefore, be in order to consider the possibility that the voters view the alternative set in some other, less structured fashion.

11.2. Results based on individual choice functions

The idea of assessing social choice procedures starting from the assumption that the voters do not necessarily have a complete and transitive preference relation over the set of alternatives but instead have an individual choice function, has recently been pursued by M. A. Aizerman and his collaborators to what seems its logical conclusion. In the following we shall briefly restate some of their results trying to explicate their implications for the theory of voting procedures (see Aizerman and Aleskerov, 1983; Aizerman, 1984).

To start with let us assume that the voters have an individual choice function C_i over the set of alternatives X so that for any subset X' of X the function C_i indicates the set X" of best alternatives in i's opinion, i.e.

$$X" = C_i(X') \subseteq X', \text{ for all } X' \subseteq X.$$

We shall consider operators F transforming the n-tuples of C_i functions into choice functions having identical properties as the individual choice functions. More formally, we shall consider the largest closing classes of operators which means that our attention is on operators satisfying the following:

if $C_i(X) \in G$ for all i in N, then

$$F(C_1(X), \ldots, C_n(X)) \in G$$

where G is a subset of all possible choice functions.

Let us furthermore restrict our focus on choice functions - individual and social - that satisfy
1) locality,
2) sovereignty,
3) monotonicity,
4) neutrality and
5) anonymity.

These are defined for choice functions in an analogous way as in the context of choice rules based on individual preference orders. The first property, locality, requires that if in two choice situations concerning the same alternative set X, x belongs to the choice set of precisely the same voter set in each situation, then it must belong to the social choice set either in both situations or in neither, i.e. if for all $X' \subseteq X$ and $x \in X'$:

$x \in C_i(X')$ if and only if $x \in C'_i(X')$, for all i,

implies that $x \in F(C_1(X'), \ldots, C_n(X'))$

iff $x \in F(C'_1(X'), \ldots, C'_n(X'))$,

then F satisfies locality. In other words, whether x belongs to the social choice set is determined solely by which individuals place it in their choice sets regardless of their attitude towards the other alternatives.

Sovereignty, in turn, requires that for any alternative in any subset of X, there is a n-tuple of individual choice functions such that this particular alternative is in the social choice set and an n-tuple of individual choice functions such that this alternative is not in the social choice set.

Monotonicity is defined similarly as above (6.1.). It is a requirement that if for any X' an alternative x is in the social choice set in some choice situation and the set of voters whose individual choice set contains x is a subset of the individuals having x in their choice set in another choice situation where X' is considered, then x should also be in the social choice set in the latter situation. In other words, if x is among the winners when X' is the alternative set, then it should remain among them if the set of individuals who have x in their choice set is either the same or contains the set of individuals choosing x in the first situation as its proper subset. Thus, monotonicity requires that additional support for x should not be harmful to it.

Neutrality requires that if the voter set having x in their individual choice sets when X is considered is the same as the voter set having x' in their individual choice sets when X' is considered, then x belongs to the social choice set from X precisely when x' belongs to the social choice set from X'. In other words, the number of voters with various choice sets and that only should decide the social choice. Aizerman and Aleskerov (1983, p. 2) observe that neutrality thus understood is a combination of the independence of context and the independence of alternatives. Indeed, it appears to be of questionable plausibility; certainly a violation of it would not be alarming as the following example shows.

Example 11.6. $|N| = 3$, $X = \{a,c\}$, $X' = \{A,C\}$

		from X	from X'
Individual choice sets:	voter 1	a	A,C
	voter 2	a	A,C
	voter 3	c	C

Here voters 1 and 2 have a and A in their choice sets in situations X and X', respectively. The neutrality property would now require that if $F(C_1(X), C_2(X), C_3(X)) = \{a\}$, then $F(C_1(X'), C_2(X'), C_3(X')) = \{A\}$. In other words, by neutrality the group should choose a from X if it chooses A from X' because the set of voters having a in their individual choice sets, viz. voters 1 and 2, is identical with the voter set having A in their individual choice sets. Yet it would not seem unreasonable to require that the choice from X should be a because two out of three voters would choose a and a only from X. But at the same time it would seem entirely plausible that C (and not A) be chosen from X' because all voters have C in their individual choice sets and no other alternative is so popular.

Anonymity, finally, requires that the permutation of the labels of individuals does not change the social choice. The properties 1) - 5) characterize social choice functions. In addition to them, we shall consider three further properties which may characterize individual choice functions as well. The first is heritage, which we already touched upon in our discussion of consistency-type properties (section 8.1.). It will be recalled that heritage requires the following: for all subsets X' of X the choice from X' contains the intersection of the choice set from X and X', i.e. $C(X) \cap X' \subseteq C(X')$.

Concordance, in turn, requires that for any X_1 and X_2 the common elements chosen from X_1 and X_2 when considered separately are included in the choice set from $X_1 \cup X_2$, i.e.

$$C(X_1) \cap C(X_2) \subseteq C(X_1 \cup X_2) \text{ for all } X_1 \text{ and } X_2.$$

The independence of rejecting outcast variants (IRO), finally, states that if X_1 is such a subset of X_2 that no element of X_1 is in the choice set of X_2, then the removal of the elements of X_1 from X_2 should leave the choice set of X_2 intact, i.e.

$$X_1 \subseteq X_2 - C(X_2) \text{ implies that } C(X_2 - X_1) = C(X_2),$$

where the minus sign is interpreted in the set-theoretic fashion.

Now Aizerman's and Aleskerov's theorem says that the intersection of those social choice functions that satisfy properties 1) - 5) above, on the one hand, and the choice functions satisfying the heritage, concordance and IRO both in individual and social choices, on the other, is empty. In other words, there are no such choice functions that satisfy all the above properties.

If we do not require that IRO condition be satisfied by the social choice function, then there is exactly one social choice operator that has all the other properties mentioned above, viz. the "unanimity" operator which is defined as follows:

$$x \in F(C_1(X), \ldots, C_n(X))$$

iff $x \in C_i(X)$, for all i in N.

If, on the other hand, we would be willing to dispense with the property of concordance, then the other properties would be satisfied by a unique operator, viz. the following:

$$x \in F(C_1(X), \ldots, C_n(X))$$

iff there exists i in N such that $x \in C_i(X)$.

Another theorem of Aizerman and Aleskerov is of a nature analogous to the famous impossibility theorem of Arrow. It states that the class of those local choice functions that satisfy the heritage, concordance and IRO both in individual and social choices coincides with dictator operators, i.e. operators with the following property:

$$x \in F(C_1(X), \ldots, C_n(X))$$

iff $x \in C_i(X)$ for a fixed individual i.

A result of a more positive nature is one that states that if we are satisfied with the properties 1) - 5) and heritage only (the latter in the sense of both individual and social choices), then a class of operators exists that can be resorted to, viz. the class

USING LESS THAN PREFERENCE ORDERS

that Aizerman and Aleskerov call k-majority operators. When these operators are used all alternatives that have been named in the individual choice sets of at least k individuals are also in the social choice set.

What then are the implications of these results for the choice of voting procedures? First of all, the negative results seem to suggest that even if the assumption that the individuals have choice functions rather than preference orders over the alternative set was found to be justified, it still would not follow that a perfect procedure for aggregating individual choices could be found. Instead, results analogous to the Arrow impossibility one would be encountered. Secondly, many of the properties or conditions featured in the preceding results seem to lack some of the normative significance that characterizes analogous properties when defined for procedures based on individual preference orders. For instance, property 1), locality, does not seem intuitively compelling. Surely this is partly due to reasons which are similar to those which make the positional voting procedures seem sometimes intuitively more reasonable than the binary ones. Also neutrality seems of somewhat questionable significance for similar reasons. On the other hand, the properties of heritage, concordance and IRO seem to capture something of what we could intuitively regard as the rationality characteristics of individual and social choices.

The over-all conclusion from the preceding survey of the problems encountered in trying to ask the voters something less than their preference order over the set of alternatives seems to suggest that more problems are created than solved by asking people to reveal their top-most preferences only (with or without ordering those alternatives). On the other hand, if it can safely be assumed that people do not even have a preference order over the alternatives but yet possess an individual choice function over the set of alternatives indicating, for each subset, the best elements in the voter's view, then the impossibility results analogous to that of Arrow are impending. If the properties of heritage, concordance and IRO are regarded as rationality properties necessary to be preserved by both individual and social choices, then only dictatorial operators are available if we restrict ourselves to local operators. If, on the other hand, less stringent rationality requirements are imposed, then, of course, the normative significance of the possibility results is diminished. In short, the possibilities of finding good procedures by asking people more or less than their preference order over alternatives do not seem to be available.

CHAPTER 12

WHY IS THERE SO MUCH STABILITY AND HOW CAN WE GET MORE OF IT?

Many of the results discussed above are of a negative nature; they state that the conjunction of several intuitively desirable properties of the voting procedures is not possible. In other words, regardless of which voting procedure one resorts to there is bound to be at least some plausible property or set of properties that do not characterize the procedure. The ubiquity of strategic manipulation possibilities and the paucity of voting equilibria make Riker (1982) very suspicious about the possibility of finding any method that would reasonably well reflect the tastes or preferences of the individuals in collective choices.

As we have observed earlier there are certainly degrees in which various procedures fail on strategic criteria of goodness. Similarly, there are degrees in which various procedures satisfy other criteria of goodness. Together these observations would seem to suggest a somewhat more optimistic view of the possibilities of finding good voting procedures. Moreover, the procedures which are widely used in practice do not seem to behave in a chaotic fashion as would seem to be implied by the paucity of equilibria alluded to by Riker.

In this chapter we shall first try to outline some possible explanations for the lack of chaos in social choices. In particular, we shall focus on such institutional and theoretical features that would make the likelihood of Pareto optimal outcomes and/or the choice of Condorcet winners more likely than when these features are not taken into consideration. Thereafter, we shall focus on ways of designing the background conditions for the application of some widely used voting procedures in such a fashion that the avoidance of some theoretical pitfalls becomes possible.

12.1. Explanations of stability

As we already noticed the results of McKelvey and Schofield deal with an infinite number of alternatives and the simple majority

WHY IS THERE SO MUCH STABILITY

pairwise comparison of alternatives. Both of these assumptions can be of doubtful validity in real-life decision making bodies. Large majorities and a limited number of alternatives often occur simultaneously in those bodies. Some results based on these sometimes more realistic assumptions were already discussed in 4.3. But how can we tell of a voting result that it is a stable equilibrium result?

If by equilibrium one means the Nash equilibrium (see e.g. Dutta and Pattanaik, 1978) or a similar outcome, the answer requires information about the individual preferences of voters. But obviously these are not always available. Indeed, many procedures currently used do not require that the voters report their full preference relations over the alternatives. Instead, a very partial expression of the relation is usually required. Let us consider the amendment procedure, which is widely used in contemporary legislatures, and the following example (see also Riker, 1982, pp. 69-73; Nurmi, 1983c).

Example 12.1.

A 100-member parliament, where parties A, B and C have 30, 30 and 40 seats respectively. The preference profile R over X = {x,y,z} is the following:

A	B	C
x	y	z
y	z	x
z	x	y

Consider another profile R' in the same parliament over Y = {a,b,c}:

A	B	C
a	c	b
b	b	c
c	a	a

Suppose now that the agenda in the first case is: 1. x vs. y, 2. the winner of 1. vs. z, and in the second case: 1. a vs. b, 2. the winner of 1. vs. c. Hence alternatives z and c are so-called status quo alternatives in the respective situations.

Now if each party in toto votes according to its preferences in both stages, the winners in situations R and R' are z and b, respectively. In both cases we have a unique winner. The procedure does not make any difference between these cases. But obviously they are markedly different. In R' we have a "genuine" winner, viz. the Condorcet winner b, whereas situation R exhibits the Condorcet paradox: x beats y, z beats x, y beats z. In R the fact that z is the winner follows from its position in the agenda, while in R' b would win regardless of the agenda.

Intuitively situation R' would lead to a more stable outcome than situation R when the amendment procedure is resorted to. This, however, is not to say that either of these situations is an equilibrium if the latter is defined in the customary way: a profile R^1 is a Nash equilibrium iff no coalition can benefit from reporting anything else than the preference it reports in R^1. Obviously by reporting the preference order yxz party A benefits in R because then y wins the first and the second ballot. Consequently, it becomes the winner instead of z, the alternative ranked worst by party A. Similarly, in R' party B benefits from reporting the preference order cab whereupon a wins in the first ballot and is defeated by c in the second one. c, in turn, is regarded as best by party B.

Even if b does not result from an equilibrium profile, it is clearly more "stable" than z when the sincere preference profile is R. But the crux is that the amendment procedure does not give us any clue to this basic difference between situations R and R'. This suggests that one explanation for the "stability" of the political outcomes is that our procedures do not tell us whether the outcomes we end up with are stable. In other words, the fact that an outcome *could* ensue from a group decision procedure as a stable outcome, given that the preference profile is R', does not entail that all outcomes resulting from the procedure would eo ipso be stable. To determine whether the outcome is stable in the sense of being a Condorcet winner, we need to know either the entire preference profile of the voters or the pairwise comparison matrix. The amendment procedure per se gives us neither: it consists of k-1 pairwise comparisons, whereas one would need up to k(k-1)/2 paired comparisons to determine the possible Condorcet winner.

Thus, one "explanation" of the apparent stability of the outcomes is that they are not necessarily stable at all. The amendment procedure simply does not make any difference between stable and unstable outcomes. This feature is not peculiar to the amendment procedure. The widely used plurality and plurality runoff procedures also share this property. The reader is referred to Table 5.1. which reports Merrill's computer simulation results on the agreement of electoral outcomes with the Condorcet winners when various procedures are used.

Another way of accounting for the apparent stability of the real world political outcomes is to look at the institutional frameworks in which the issues are prepared for eventual social choice. This is the approach pursued by Shepsle (1979) with interesting results (see also Shepsle and Weingast, 1982). Specifically, Shepsle argues that the notion of structure-induced equilibrium makes it plausible to maintain that the real world outcomes are often equilibria in this specific sense even though they might not be so when viewed from the angle of multidimensional spatial models. In other words, even though the observed outcomes are not stable in the sense discussed in spatial theory, they are stable in the sense that given the structural constraints they cannot be upset.

A special and rather untypical case of such an equilibrium would be an outcome resulting from a process that is straight-forward in the following sense: the dominant strategies of the voters are sincere preference revelations. An example of a straight-forward procedure is discussed by Fishburn (1983, p. 389): an odd number of voters decides about the amount of funds to be allocated to a program. We assume that each voter has an ideal amount and that the further away from it the outcome is, the less pleased he is. The decision is made by closed ballot whereby each voter writes an amount on a slip of paper. The median amount is then declared the winner or social choice. This procedure is straight-forward as no voter can benefit from voting for another amount than his ideal. If, on the other hand, the social choice would be determined as the arithmetic mean of the individual ballots, the sincere revelation would no longer be dominant and the process no longer straight-forward. The point, however, is that straight-forward processes or less stringent structure-induced equilibria may occur in the real world when the alternatives are considered dimension by dimension and not in all dimensions simultaneously.

Another institutional feature that certainly excludes the "wandering" of the outcome sequences outside the Pareto set is the requirement that the agenda be built according to specific rules. Consequently the agenda-setter's powers are limited in real life decision making bodies (see e.g. Niemi, 1983; Nurmi, 1980). For example, the speaker's rules in the Finnish parliament require that when a vote is taken over a set of alternatives the first ballot is taken between the two alternatives farthest apart. From these alternatives one then proceeds towards the center. Thereby the centrist alternatives have a definite advantage if the voting is sincere as has been shown by Black (1958) (see also Fishburn, 1983).

The explanations of stability considered thus far tend to lean on the fact that real world outcomes are not necessarily stable after all - hence the explanandum evaporates - or the procedures utilized in specific institutional settings make them stable relative to those settings. We now turn to the possibility of seeking the explanation of this stability in the models themselves.

One feature of the spatial models underlying the instability results is that whenever an alternative x is closer - in terms of a given norm - than y in the policy space, the voter is assumed to vote for x. In the probabilistic voting models of Hinich et al. (1973) this assumption is weakened. Consider a two-candidate contest in the Euclidean m-space X where the location of candidate 1 is the variable c_1 and that of candidate 2 the variable c_2. The optima of voters $1, \ldots, n$ are denoted by $v_1, \ldots, v_n$, respectively. Probabilistic voting is based on three main assumptions:

1) the continuity and concavity of $U_i(x| v_i)$ over X, where $x \in X$ and $U_i(x| v_i)$ denotes the utility of alternative x for voter i whose optimum is at v_i. Continuity means that the function U_i is defined everywhere in X and its changes in any direction from v_i are smooth. Concavity, in turn, means that the further away from v_i the point x lies, the smaller its utility for i and, moreover, for any x,y in X the following holds:

$U_i(px + (1-p)y| v_i) \geq pU_i(x| v_i) + (1-p)U_i(y| v_i)$, where $0 < p < 1$.

Furthermore, the utility decreases by a constant or increasing rate with the distance from v_i.

Let now $p_1^i (U_i(c_1| v_i), U_i(c_2| v_i))$ and $p_2^i (U_i(c_1| v_i), U_i(c_2| v_i))$ be the probability that i votes for candidate 1 and candidate 2, respectively. Then

2) p_1^i is strictly increasing with the increase of $U_i(c_1| v_i)$ and p_2^i similarly strictly increasing with the increase of $U_i(c_2| v_i)$. Finally,

3) p_1^i is a monotonically decreasing function of $U_i(c_2| v_i)$ and p_2^i similarly a monotonically decreasing function of $U_i(c_1| v_i)$.

Now under these (and an additional technical assumption concerning the integrability of p_1^i and p_2^i with respect to c_i and c_2) Coughlin (1982) shows that whenever c_1 takes on a fixed value c'_1, a position of c_2 which aims at maximizing the expected number of votes against c'_1, or the expected plurality against c'_1, must be in the Pareto set. This implies that the convergence of the outcomes on the Pareto set is very swift, indeed. Moreover, the expected vote-maximizing trajectory no longer departs from the Pareto set once it has entered it.

These assumptions relate to those of deterministic voting so that assumptions 2) and 3) are replaced in deterministic voting by the following:

2') $p_1^i = 1$ if $U_i(c_1|\ v_i) > U_i(c_2|\ v_i)$,
$p_2^i = 1$ if $U_i(c_1|\ v_i) < U_i(c_2\ |\ v_i)$ and
if $U_i(c_1|\ v_i) = U_i(c_2|\ v_i)$, then i abstains.

The probabilistic voting allows for all kinds of non-spatial considerations to enter the voter's voting calculi. For instance, the views of the candidates' personalities might well make it plausible to assume that voter i has a non-zero probability of voting for candidate 1 even though $U_i(c_2|\ v_i) > U_i(c_1|\ v_i)$. Indeed, this probability could even reflect the possibility that the voter by accident pushes the wrong button or writes the wrong number in the ballot booth. Nevertheless, probabilistic voting is not a generalization of deterministic voting because the latter by violating condition 2) is not a special case of the former. Despite this, the probabilistic considerations seem plausible enough and would certainly account for the stability of real world political outcomes at least insofar as the outcomes tend to be Pareto optimal.

Enelow and Hinich (1984) come up with another intuitively plausible explanation of the fact that "centrist" outcomes are rather common, intuitively speaking. The explanation is again in terms of a probabilistic voting model which renders the centrist outcomes stable. Assuming probabilistic voting which allows for abstention due to alienation and indifference, Enelow and Hinich show that such an equilibrium point exists that guarantees the expected plurality of votes for a candidate located at this point in what they call predictive space, i.e. a space of fundamental policy issues which the voters are using in inferring the candidates' policies on more specific issues.

Two additional details make the result particularly interesting for our purposes:

i. a ranking of all policy positions in the predictive space can be established on the basis of the expected plurality of votes, i.e. position x is ranked higher than position y iff the expected plurality for a candidate occupying x when confronted with a candidate occupying y, is greater than zero, and

ii. optimal position is a weighted mean of the voter's optima in the predictive space. It is, however, not the case that the optimal position is obtained by simply using the group sizes as weights. In addition to the size, other considerations like the subjective policy significance of the predictive dimension and the importance of alienation and indifference play important roles in determining the equilibrium outcome. The point is, however, that centrist positions are likely to be equilibrium ones, a result that simply does not hold for deterministic voting in multidimensional spaces as the instability results show.

So it would seem plausible to argue that the instability results assume a kind of "perfect" if also myopic voting behaviour that does not characterize real world voters. But an even stronger claim can be made: the deterministic voting results are not only descriptively but also normatively questionable as they give a very restricted role to abstentions due to alienation and indifference and assume such a short-sightedness among voters that cannot reasonably be called rationality. Let us finally take a closer look at the latter assumption.

The arbitrariness of the simple majority rule in the absence of a core is the main message of McKelvey's theorems. However, the theorems are based on a behavioural assumption that may not be entirely plausible. The agenda may consist of consecutive alternatives that are very far apart. To assume that each voter always votes for the alternative which happens to be closer to his optimum without any regard for the larger portion of the agenda, can only be justified if the agenda in toto is not known to the voters. In some cases the arbitrariness of the majority rule may involve violent moves across the Pareto set. It would seem rather strange if the voters would go along with these kinds of moves without smelling a rat. Rather, one would expect them to vote in a fashion that is myopically irrational, i.e. voting for x in comparison with y even though $u_i(x) < u_i(y)$ or then abstaining if the corresponding utilities are almost but not quite equal.

The relevance of Schofield's results on the genericity of *local* cycles in higher dimensions is precisely in their demonstration that the arbitrariness of the simple majority rule holds also locally in higher dimensional voting games. But then one can point to the counterintuitiveness of assuming that the voters respond to very small - indeed infinitesimal - utility differences between alternatives. Rather it would seem plausible to assume that rational voters can vote against their short-term preferences in order to gain in long-term ones. This is not to play down the theoretical importance of McKelvey's and Schofield's results. They are certainly convincing in showing the arbitrariness of the simple majority rule. What seems worth looking into is whether a somewhat more intuitive voter model would explain the stability or central tendency of the political outcomes.

The concept of nonmyopic equilibrium has been introduced by Brams and Wittman (1981) for 2 by 2 games (see also Brams 1983). Consider the following game:

WHY IS THERE SO MUCH STABILITY

		Column	
		C	D
Row	A	R_{11}, C_{11}	R_{12}, C_{12}
	B	R_{21}, C_{21}	R_{22}, C_{22}

Suppose that the game is played sequentially, i.e. starting from any cell R_{ij}, C_{ij} in the matrix the players can with alternating moves shift the outcome away from that cell so that Row can move the outcome to the cell $R_{jj}C_{jj}$. Similarly, Column can change the outcome from $R_{jj}C_{jj}$ to $R_{ji}C_{ji}$, where i,j = 1,2. The sequential game has reached a terminal outcome iff the player who is in turn, say Row, cannot benefit from a move because the current outcome R'_{ij}, C'_{ij} is such that $R'_{ij} = \max_{ij} R_{ij}$.

Now a nonmyopic equilibrium is defined as follows. An outcome R'_{ij}, C'_{ij} is a nonmyopic equilibrium iff it is a nonmyopic equilibrium for Row and Column. R'_{ij}, C'_{ij} is a nonmyopic equilibrium for Row (Column, respectively) iff all continuations of the game with alternating moves result in terminal outcomes R_{ij}, C_{ij} such that $R_{ij} < R'_{ij}$ ($C_{ij} < C'_{ij}$) and moreover R'_{ij}, C'_{ij} is the surviving outcome in the process of backward induction proceeding from the terminal outcome to R'_{ij}, C'_{ij}. The process of backward induction is illustrated in the following game-tree of the Prisoner's Dilemma (the matrix form is presented on the left).

Example 12.2.

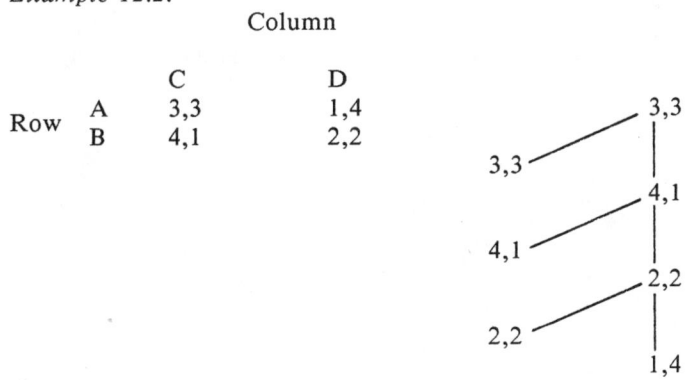

Starting from the terminal outcome 1,4 we compare it with 2,2 in terms of Row's payoffs. Obviously, 2,2, is better and therefore survives this comparison. 2,2, is then compared with 4,1

in terms of Column's payoffs. Again 2,2 survives. Finally 2,2, is compared with 3,3 in terms of Row's payoffs. The latter clearly survives. Thus starting from 3,3 Row cannot guarantee himself a higher payoff than 3,3. Hence 3,3 is a nonmyopic equilibrium for Row. Similarly 3,3 can be seen to be a nonmyopic equilibrium for Column. It is therefore a nonmyopic equilibrium in the Prisoner's Dilemma.

A similar idea has been used in the study of binary voting games. Perhaps the earliest account is Farquharson's (1969). He introduces the notion of "sophisticated voting". This concept is defined in terms of the domination relation between voting strategies. Farquharson considers a process whereby the voters successively eliminate all but undominated strategies. If the process is determinate, a unique equilibrium outcome ensues. This seems to capture at least some aspects of nonmyopic behaviour. Farquharson's procedure is, however, rather cumbersome to apply and we shall, therefore, focus on a multistage game representation method devised by McKelvey and Niemi (1978). Consider the following binary voting game.

Example 12.3.   X = {a,b,c,d,e}   N = {1,2,3}

person 1	person 2	person 3
a	b	c
b	a	d
c	e	a
d	d	b
e	c	e

The procedure utilized is the successive procedure. Suppose that we have the following agenda:

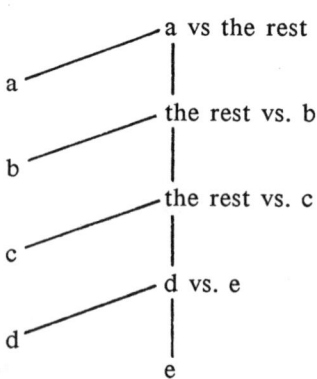

If everyone votes sincerely at each stage, b will be chosen. For each node of the game-tree McKelvey and Niemi define the sophisticated equivalence of the node as follows:

(i) if the node is terminal, then the sophisticated equivalence (SE, for short) is simply the node itself, and

(ii) if the node is nonterminal, its SE is defined on the basis of the social preference between the two alternatives (branches) emanating from it. Thus supposing that the simple majority is used in all paired comparisons to determine the social preference, we get the following tree of SE's proceeding from the terminal nodes towards the top of the tree:

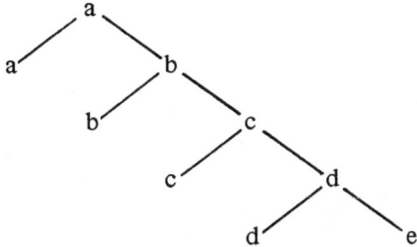

The concept of a sophisticated voting strategy now indicates the way in which the voter votes at each node of the game-tree: he votes for that branch which represents the SE preferred by him to the other. In our example the sophisticated voting strategies for persons 1, 2 and 3 are:

$$s_1 = (0,0,0,0)$$
$$s_2 = (1,0,1,1)$$
$$s_3 = (0,1,0,0)$$

where 0 indicates the choice of the left branch and 1 the choice of the right one. Thus the sophisticated outcome - the outcome resulting from the choice of sophisticated strategies - is a, the Condorcet winner. The choice of the Condorcet winner is not accidental; McKelvey and Niemi show that the Condorcet winner will always be chosen by sophisticated voting strategies when binary procedures are resorted to. Another result by the same authors states that when a top cycle exists the outcome resulting from sophisticated voting is the choice of an alternative from that cycle.

Could the sophisticated voting in the above sense, then, be an explanation of the stability of the outcomes? In a sense yes, viz. if the increase in the likelihood of the choice of the possible Condorcet winner is what is meant by increasing stability. Consider the following application of the amendment procedure:

Example 12.4.

The preference profile:

person 1	person 2	person 3
v	z	y
x	v	z
y	x	v
z	y	x

The agenda:

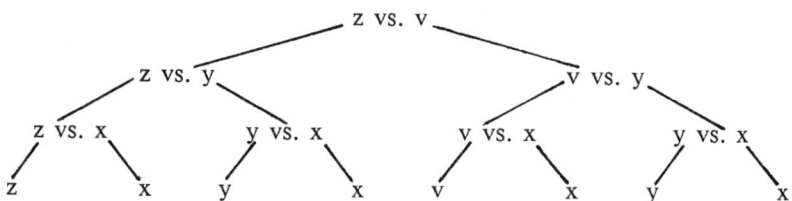

By backward induction we get the following tree of SE's:

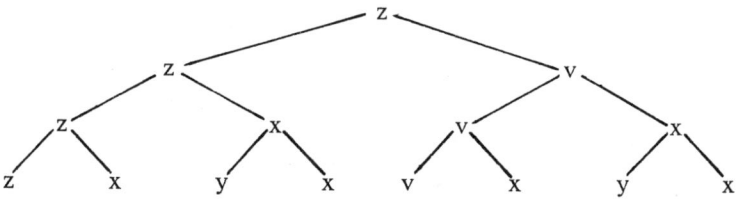

Now obviously z becomes the sophisticated voting winner, whereas with sincere voting and the same agenda we would end up with Pareto dominated x. So at least in this example the sophisticated voting would avoid the worst pitfalls of the amendment procedure and so in a way explain the centrist tendency of the real world outcomes.

Also in the following case the sophisticated voting equilibrium belongs to the Pareto set whereas with the same agenda the sincere voting would lead to a Pareto dominated outcome (Riker, 1982, p. 74).

WHY IS THERE SO MUCH STABILITY

Example 12.5.

Consider the alternative set X = {x,y,s,t,w} and the agenda:

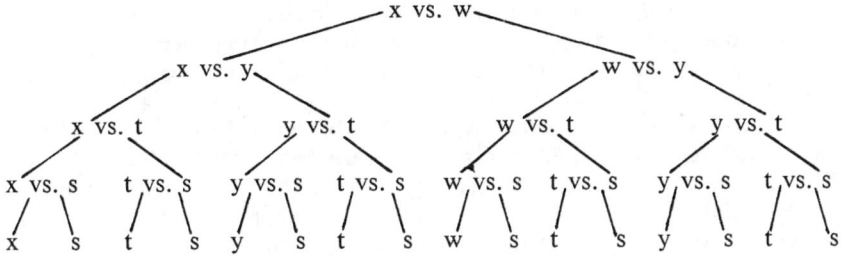

Suppose that the individual preferences are as follows:

person 1	person 2	person 3
w	y	s
x	w	x
t	x	t
y	t	y
s	s	w

In this example the tree of SE's is the following:

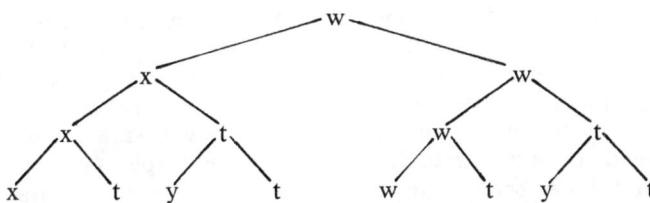

The sincere voting would result in the choice of t which is Pareto dominated by x. The sophisticated voting outcome, in turn, is w, a Pareto undominated alternative.

No finite number of examples will, however, prove the general theorem that the sophisticated voting never results in a Pareto dominated alternative. However, looking at McKelvey's and Niemi's theorem and its two corollaries it turns out that whenever a Condorcet winner or a top cycle exists, sophisticated voting leads to a Pareto optimal outcome. These results imply that in general the possibility of a Pareto dominated choice is very limited. Indeed, if all preferences are strict, it is excluded.

If by stability is meant the necessary choice of the Condorcet winner (when one exists) and the Pareto undominance of outcomes, we have touched upon above a number of considerations that would explain why stable outcomes have occurred. On the other hand, none of these explanations makes stability necessary in the real world. The models discussed can also accommodate unstable outcomes. Even though the moral issues have not been taken into consideration in the preceding one could well ask if the search for stability would undermine other values that we hold dear. This has actually been argued by Miller (1983). He maintains that restrictions on the admissible preferences of individuals are efficient ways of guaranteeing the transitivity of the social preference relation and other pleasant properties. However, these restrictions obviously contradict the very rationale of using individual preferences as the point of departure in social choice. As we have seen this argument - although plausible if deterministic and myopic voting is assumed - loses some of its plausibility when the above "relaxed" models are focused upon.

12.2. Improving the performance of the voting procedures

We already observed in the preceding section that the users of the voting procedures have found it useful to institute various auxiliary guidelines or rulings that seem to improve the performance of the amendment procedure. Also in the context of the plurality runoff system we noticed the prohibition on the publication of opinion polls for some limited period preceding the elections. The former rulings seem to be motivated by the desire to limit the agenda-manipulation possibilities of the agenda-setters, while the latter aims at giving people incentives to reveal their true preferences instead of the strategic ones. Both examples pertain to the features that have been found by Riker to make the entire idea of aggregating individual wills futile, viz. the ubiquity of the strategic manipulation possibilities and the omnipresence of majority cycles or other nasty phenomena. The thought of patching up other procedures as well would then seem worth pursuing, as obviously the results by Shepsle briefly outlined in the preceding section show that if we overlook the institutional surroundings of the voting procedures we might miss something important in their behaviour.

We have already paid attention to the fact that speaking of manipulability in the sense of individual preference misrepresentation is likely to blur the issue to the extent that this property is considered as a dichotomous one. Even though all the procedures are manipulable in this sense, there are essential differences between them concerning the difficulty of manipulation or the amount of information one has to possess about other people's preferences in order to benefit from preference misrepresentation. E.g. Hare's procedure is extremely difficult to manipulate and some empirical evidence gives strong support to this claim (see Chamberlin et al., 1984).

But suppose one has to utilize a procedure that is easily manipulable in this sense, for example the Borda count. How could one make its manipulation more difficult? Obviously by restricting the amount of information available for each voter concerning the preferences of others. The prohibition on the publication of opinion polls would do the trick for mass elections but in small groups of people where everybody knows the preferences of all the others or nearly so, the use of the Borda count would clearly create incentives for preference misrepresentation.

Similarly, Coombs' procedure would easily lead to the same type of manipulation, viz. giving the likely winner or sincere winner lower rank in one's preference order than one sincerely would deem appropriate. But in small groups one could more efficiently than in large ones appeal to group norms in counteracting preference misrepresentation. For example, one could try to convince the voters that the misrepresentation of preferences would in a way undermine much of the rationale of resorting to group choice in the first place. Surely this kind of argumentation would only work under circumstances where the individuals are sufficiently disinterested in the social choices. With sufficiently high stakes it is probable that anyone would misrepresent his preferences if the chances of success were good enough. It is impossible to get a person to play a game "fairly" or according to agreed upon rules, if the person is convinced that what the game is about is not "fair", e.g. the outcomes affect the participants in a disproportionate manner.

One way of invoking the group norms to counteract the incentive for preference misrepresentation would be to make every individual aware of the likely reactions that preference misrepresentation would cause in other persons, supposing that there is a way to determine that some persons are misrepresenting their preferences. One would then simply be instituting a Prisoner's Dilemma norm in order to avoid the group being locked in a non-cooperative equilibrium in a collective action game (see Ullman-Margalit, 1977). But if sincere preference revelation represents the cooperative strategy and the preference misrepresentation the non-cooperative one and the payoff matrix is that of the Prisoner' Dilemma, then the enforcement of the cooperative choice by all voters will produce - in addition to the usual problems of ensuring cooperation in the Prisoner's Dilemma - the problem of determining whether an individual has defected to the non-cooperative strategy. To determine whether this has occurred one would need independent information about the sincere preferences of the voters.

Of course it would be much easier to resort to procedures that are more difficult to manipulate, regardless of the size of the group. But preference misrepresentation is but one form of strategic manipulation as we observed above. The problem of agenda-manipulation is equally serious even though as we just noticed some attention has been paid to it in real world decision making bodies. The power of the agenda-setter is essentially diminished if there is an agreed-upon method of constructing the agenda so that whenever a social choice procedure is to be applied there is only one way of constructing the agenda and the alternative set. In practice a perfect set of rulings would be difficult to devise. For example, even though the principle of subjecting those two alternatives that are farthest apart to the pairwise comparison first, solves some of the problems related to the amendment method it is often impossible to agree on the policy space and the norm to be used in measuring the distances of the alternatives.

But for the binary procedures it would of course be helpful if the input data used in determining the winners were the preference orders of the voters instead of the distribution of votes in pairwise comparisons. This would immediately give the persons in charge of the computation of the voting results the data for all $k(k-1)$ paired comparisons. It would do away with the problem of fiddling with the order of voting in order to end up with suitable outcomes. But it would still leave the persons deciding, which alternative set is to be considered, a considerable power to manipulate the outcomes. Almost the only possibility of getting around this problem is to give all members of the collectivity the power to suggest new alternatives to the agenda. This would not eliminate the problem but would make the members of the body equal in the manipulation of possibilities.

Looking at the manipulability problematique from a slightly different angle one could ask whether we are in the end dealing with a minor problem as far as the misrepresentation of individual preferences is concerned. After all there is no such thing as a "pure" individual preference order if by that one means an order that is unaffected by other persons' preferences. Are we then not dealing with just another form of taking the preferences of others into account in one's own preferences? What is the point of regarding one type of influence as unacceptable while no precautions

WHY IS THERE SO MUCH STABILITY

are taken to stop people from modifying each other's preferences by other means? These questions are well worth asking even though one's attitude towards them is to some extent determined by considerations beyond the scope of this study.

The preceding discussion shows that the procedures differ widely with respect to properties other than the strategic ones. One is thus led to ask whether there are means of improving their performance that have not been taken into account in the preceding. Combining the procedures would prima facie seem a good way of trying to preserve the desirable properties of the systems, but as we shall show shortly this may lead to quite counterintuitive results.

The best way to start would be to ask the voters to give their entire preference orders so that the actual occurrences of failures on various criteria can be determined when any given procedure is used in preference aggregation. For instance, if approval voting were used the voters could reveal their entire preference orders indicating also the set of alternatives they approve of. Using the data thus obtained one could determine the number of Condorcet criteria and/or Pareto optimality violations. Moreover, in the case in which the procedure utilized would lead to a Pareto dominated outcome, some of the alternatives unanimously preferred to it could be chosen e.g. randomly. Similarly should the procedure actually used lead to the choice of the Condorcet loser, one could eliminate this alternative from the choice set when the preference profile in toto is known.

The determination of the occurrences of Condorcet criteria violations as well as of the choice of a Pareto dominated alternative is relatively straight-forward: for a fixed preference profile and alternative set, these phenomena either happen or do not happen, assuming that the procedure utilized makes their occurrence possible. Things are more complicated with respect to monotonicity and choice set invariance criteria.

The failure of monotonicity means that, had the actual winner gained more support than it did, it would not have been chosen. The violation of WARP would e.g. mean that even though x won when X was the alternative set, it would not have done so, had the alternative set been some proper subset of X containing x. These violations seem to involve a counterfactual element: had the choice situation been different from what it actually was, then the choice would have been different as well.

As was argued earlier, monotonicity seems a quite essential theoretical property of democratic procedures and, consequently, its violations seem particularly serious for the entire enterprise of social choice. If monotonicity violation "occurs" in the sense that it can be shown that additional support for the winning alternative would have been fatal for it, it certainly seems to undermine the legitimacy of the chosen alternative.

But, on the other hand, many non-monotonic procedures e.g. Hare's and Coombs' methods, behave in a plausible fashion under a wide class of preference profiles. For instance, there are no problems when an alternative with more than 50 % of the first ranks exists. But also in more general settings these procedures do not necessarily involve monotonicity violations. However, the only way to determine when they do so, is to ask for the entire preference orders from voters. How to react to the eventual monotonicity failures is then something to be decided upon separately.

The same, of course, applies to the violations of choice set invariance criteria even though in their case, if one wants to exclude these failures altogether, the remaining procedure repertoire is rather small. Anyway, instead of concluding that the procedures currently used are such as to make the social choices meaningless, perhaps one should look at the issue in the light of the frequency of the violations of desirable theoretical properties. After all what is possible is not ipso facto common and certainly not necessary in all cases.

When looking at the comparative performance of various procedures in the light of the criteria we have resorted to above, one might like to play with the idea of combining the various procedures so as to end up with "hybrid" ones in order to overcome the various weak points of the constituent procedures. For instance, when glancing through the comparison of the voting procedures one might be tempted to combine the Condorcet winning criterion with approval voting so as to enhance the performance of the resulting procedure. However, if one constructs a procedure that works so as to choose the Condorcet winner whenever such an alternative exists in the preference profile reported by the voters and otherwise chooses the approval voting winner on the basis of the information given by the voters concerning the acceptable alternatives, then one has the benefit of satisfying the Condorcet winning criterion by fiat, but one simultaneously loses heritage and thus WARP and path independence, as the following example shows.

Example 12.6. $|N| = 3$, $X = \{a,b,c\}$

voter 1	voter 2	voter 3
a	b	c
b	c	a
c	a	b

WHY IS THERE SO MUCH STABILITY

Suppose that voters 1 and 3 approve of their first ranked alternative only, while voter 2 gives his approval to both b and c. Then the above "hybrid" procedure would result in the choice of c. However, c is not chosen by the same procedure from the subset {b,c} where trivially b is the Condorcet winner. Hence heritage is violated. The procedure might also be inconsistent as the following example illustrates.

Example 12.7. $|N| = 13$, $X = \{a,b,c\}$

The preference profile of group N_1:

voters 1-3	voters 4-5	voters 6-7
a	b	c
b	c	a
c	a	b

The preference profile of group N_2:

voters 8-9	voters 10-11	voters 12-13
c	b	a
a	c	c
b	a	b

Now in group N_1 there is no Condorcet winner. Suppose that voters 4 and 5 approve their first two alternatives, whereas the other voters in N_1 approve their first ranked alternatives only. Then the choice of N_1 is c. The same alternative c is the Condorcet winner in N_2 and is, thus, chosen. However, when the groups are combined c is not necessarily chosen any more. In $N_1 \cup N_2$ there is no Condorcet winner. If now voters 8 and 9 approve both c and a, whereas the others in group N_2 approve their first ranked alternatives only, the choice of $N_1 \cup N_2$ is a. This violates consistency.

The combination of procedures is, thus, not per se a solution to problems of finding optimal voting procedures. Whether the hybrid procedures satisfy the various criteria of goodness remains a contingent matter not directly inferable from the properties of the constituent systems.

CHAPTER 13

FROM COMMITTEES TO ELECTIONS

So far we have not touched upon the nature of the alternatives from which the choice is to be made. No difference has been made between cases in which the alternatives are candidates to offices and cases in which they are policy alternatives to be implemented. The reason for not making this distinction is simply that the problems discussed above are encountered in both cases. We are dealing with "direct democracy" in both cases. Of course when we are choosing representatives to make policy choices on our behalf, then the connection between our policy preferences and the actual policy choices by representatives becomes an indirect one, but choosing representatives and policies is otherwise basically the same kind of activity. One difference is, however, worth making, viz. the fact that when representatives are elected, we are often dealing with multimember bodies while policy choices usually result in one alternative. Hence, new types of problems are encountered in trying to guarantee that the elected body represents the electorate at large in a reasonable way. In the following we shall focus on these problems which are peculiar to electing representative bodies.

13.1. Proportional and majoritarian systems

The variation among actually employed electoral systems is vast; at least the following categories can be found: plurality, majority, semi-proportional and proportional ones (see Bogdanor, 1983, p. 17). However, from the theoretical view-point the most interesting and also the most basic distinction differentiates the majoritarian and related systems from the proportional ones. These two system types can be thought of as motivated by essentially different views of representation.

The majoritarian systems - among which I shall also count the plurality systems - do not primarily strive to make the elected body as similar as possible to the electorate at large with respect to views, tastes or preferences. As Bogdanor (1983, pp. 6-7) puts it: ". . . majority and plurality systems share one fundamental feature: the number of seats which a party receives depends not only upon the number of votes which it gains, but upon where these votes are located."

On the other hand, the proportional systems are based on an effort to have the relevant views distributed in a roughly identical fashion in the elected body and in the electorate. From this angle the majoritarian methods can be seen as limiting cases of proportionality: in one-member constituency systems there is no consideration of proportionality, while in k-member constituency (i.e. one-district) systems one is as near to perfect proportionality as one can get if the number of representatives is k. The crux of the distinction would, then, seem to be both the number of representatives elected from a constituency and the total number of constituencies.

But surely this would leave open the issue of which one of a number of methods or criteria to use in determining which candidates have been elected once the individual ballots have been cast. The issue here is not primarily the differences between e.g. d'Hondt's or Sainte Lagüe's proportionality formulae, but a more fundamental or conceptual issue, viz. what we really mean by proportionality. The idea of choosing candidates in proportion to the amount of support given to them by the voters can be given different interpretations as the following example illustrates.

Example 13.1.

Suppose that in a two-member constituency there are four candidates running and that the voters can be divided into four groups so that all members in each group have identical preference orderings over the candidate set. The groups I-IV comprise, respectively, 30, 30, 25 and 15 voters of the 100-strong electorate in the constituency.

I	II	III	IV
a	c	d	b
d	b	b	d
b	d	c	a
c	a	a	c

Supposing that no parties or alliances have been formed, there are several ways of allocating the two seats to the candidates so that the allocation is "proportional" in some sense. The most straightforward of them is one which gives both a and c a seat as these candidates have been ranked first by most voters. On the other hand, another concept of support would give the seats to b and d as these have the largest Borda scores.

In this example the single transferable vote system which specifically aims at proportionality gives the following result. No candidate exceeds the quota, which is 34 i.e. 100/3 + 1 rounded down to the largest integer. The alternative with the smallest number of first ranks, b, is now eliminated and the votes of group IV are transferred to its second ranked alternative, d. This alternative now has 40 first ranks thus exceeding the quota. The votes of group III and IV are now transferred to c and a, respectively. Both of course now exceed the quota, but since only one of them can be chosen it will be c. Thus the single transferable vote system results in c and d.

We have thus three different results by giving different interpretations to proportionality. The reason is obvious: the concepts of proportionality employed are based on a different underlying social choice function, the result b, d on the Borda count and the result c, d on the single transferable vote. This example shows that even when the system resorted to is "proportional" we are dealing with basically similar problems in elections as in committees. It also shows that without further specification the notion of proportionality is essentially ambiguous.

The role of parties and electoral alliances is, of course, marked in many contemporary electoral systems. Party lists, for example, modify the choice situation of the voters considerably. We shall, however, not dwell on specific systems, but focus on more general characteristics and criteria. The majoritarian - i.e. plurality and majority - procedures are widely applied in single-member constituency systems. Their virtues are more obvious when the geographic representation is deemed important and when the parties do not feature as the main foci of the ideological identification of voters. The candidate with the majority or plurality of votes can, indeed, be regarded as the representative of the entire constituency.

On the other hand, the notorious problem of the "wasted votes", i.e. votes given to the non-winning candidates in various constituencies, is unsolved in majoritarian systems. It is well-known that in plurality systems the wasted votes can add up to more than 50 % of the total votes given in the entire system. Since single-member constituencies and majoritarian systems are straight-forward applications of plurality or plurality runoff type procedures we shall refrain from further analysis of these systems. Instead we shall move on to consider those aspects of proportional representation systems that are not immediately obvious extensions of the ideas we have touched upon in the preceding chapters.

13.2. Criteria for proportional systems

Proportionality considerations pertain to two phases of the electoral process:
1) the allocation of seats to the constituencies, and
2) the allocation of seats to candidates within the constituencies.

The first phase is usually called the apportionment. Curiously enough in systems where proportionality is explicitly aimed at, the methods of implementing it may vary in phases 1) and 2). Usually in the apportionment phase there are very few other considerations apart from proportionality to guide the allocation principles, whereas in phase 2) there often are other goals that interfere with proportionality-maximization, such as the achievement of clear majorities in parliaments to make the formation of majority governments easier.

In many systems the apportionment formula is

$$\frac{s_i}{s} = \frac{n_i}{n}$$

where s_i is the number of seats allocated to constituency i, s is the total number of seats in the body, n_i is the population of constituency i and n is the size of the total population. The problem is that s_i is typically not an integer. What is one to do with the fractional remainders as obviously by giving each constituency i the integer part of s_i leads to a situation where the sum of allocated seats is less than s?

The method normally utilized in systems where the above formula is used in apportionment is to give each constituency i the number of seats indicated by the integer part of s_i and then proceed to allocate the remaining seats to constituencies according to the size of the fractional remainders. Thus the constituency j with the largest fractional remainder gets the integer part of s_j plus one seat, the constituency 1 with the second largest remainder gets the integer part of s_1 plus one seats etc. until the sum of allocated seats is s.

The method is incompatible with one of the criteria devised for the evaluation of proportional representation systems, viz. monotonicity. This criterion can further be subdivided into
    a) house-monotonicity, and
    b) population-monotonicity (see Balinski and Young, 1982, pp. 43-44; Carter, 1982).

No doubt the most famous violation of a) is the Alabama paradox in the electoral history of the United States (see also Brams, 1976, pp. 137-166). The problem emerged in deliberations concerning the apportionment of seats of the US House of Representatives to states when the total number of seats in the House varies. The Alabama paradox involves the finding that according to Hamilton's formula then used in the apportionment (the formula just defined above), Alabama was entitled to 8 seats out of 299 but to only 7 had the total number been 300. That this phenomenon was found unsatisfactory led eventually to the abolition of Hamilton's method in the apportionment of the US House of Representatives. Theoretically this abolition was an expression of the requirement a) stating that if the total number of seats is increased, no constituency should be entitled to fewer seats than before.

Criterion b), in turn, amounts to the requirement that if the population of a constituency i increases relative to that of constituency j, it should not happen that i loses one or more seats to j. Hamilton's method does not guarantee that this requirement always holds.

Also a third type of non-monotonicity can be encountered when Hamilton's method is used, viz. the paradox of new states (Balinski and Young, 1982, p. 44). This phenomenon has also occurred in United States electoral history. Oklahoma entered the Union in 1907 and by Hamilton's method was given 5 seats out of the 391 in the House. Before Oklahoma joined, the total number of seats was 386, the state of Maine having 3 and New York 38 seats. With the entrance of Oklahoma with the additional 5 seats, the distribution of seats between Maine and New York changed: Maine now had 4 and New York 37. And yet, the seats to which Oklahoma was entitled were added to the total number.

If the avoidance of monotonicity paradoxes were the only consideration in designing electoral systems, one would be well-advised to adopt some kind of divisor method. E.g. d'Hondt's method, which is a divisor one, works as follows. One finds a divisor such that when the population size of each constituency is divided by it, the integer parts of the quotients add up to the total number of seats. Obviously this method cannot violate monotonicity in any of the senses discussed above. Unfortunately, another problem haunts the divisor methods, viz. quota violation. This takes on two forms:

(A) the upper quota violation, and

(B) the lower quota violation.

The former happens whenever a constituency i is given more seats than

$$\frac{n_i}{n} s + 1.$$

The latter, in turn, means that a constituency i is given fewer than

$$\frac{n_i}{n} s - 1 \text{ seats}.$$

While quota violations may occur when divisor methods are used, they cannot occur when Hamilton's method is used. The reason for the latter is obvious from the computational formula of Hamilton's allocation.

The two sets of criteria, monotonicity and quota satisfaction, thus cannot be fulfilled by either divisor or quota methods. In the preceding they have been introduced in the context of the apportionment problem, but of course analogous considerations would have been encountered had our focus been on the problem of allocating seats to parties or electoral alliances, given the support to these groupings (see Nurmi and Lagerspetz, 1984, for examples from the Finnish political system). In practice paradoxical features like non-monotonicity do not play a very important role for two reasons.

1) Even though the method is vulnerable to these undesirable phenomena, they may not occur because of other institutional arrangements. In other words, the antecedents of the "if-then" statements of the criteria may never materialize. A case in point is house-monotonicity when the size of the legislature is kept fixed. Obviously no Alabama paradox can occur in such circumstances. Similarly if the size of the voting body is fixed, there can be no new states paradox.

2) The differences in the allocation outcomes of various methods - divisor or quota - tend to be rather small. When this observation is combined with the explicitly stated goal of many political systems to deviate from perfect proportionality in order to achieve some other objective which is deemed to be of at least equal importance (like encouraging the formation of strong cabinet coalitions), the issue of which particular computational formula results in the most paradox-free allocations may be deemed to be of secondary importance.

Indeed, if one strives for near-perfect proportionality in seat distribution, the problem of which formula to use in either determining the apportionment of seats to districts or the seat distribution over the parties within each constituency, does not matter very much once it is decided that there ought to be several constituencies. The act abolishing the multiple constituencies would do away with the most significant obstacle to proportionality (Nurmi and Lagerspetz, 1984). In one-constituency systems the problem of which computation procedure to use would gain some importance. But even there it seems that some foundational questions must be answered before one can attach any meaning to proportionality. Let us now turn to these often unanswered and perhaps in the end unanswerable problems.

13.3. Voting power

The most obvious goal of proportional systems is to give various groupings in society a representation in the collective body that would more or less reflect its relative size in that society. But surely the importance of the collective body is more often than not in the influence it exerts, through its decisions, over the entire population. This is particularly obvious in legislatures: its decisions are typically binding for the entire population. But what does one then strive for when giving the groupings seats in proportion to their relative sizes?

The most natural answer would seem to be that, purely ceremonial goals notwithstanding, one wishes to give to representatives of the groupings an influence over the body that would correspond to the relative size of the groupings in the society at large. But if this is the primary aim, then even perfect proportionality which makes the relative share of seats over groupings identical with the relative size of the groupings in the society, falls short if the influence over the acts of the collective body is measured by the common indices of a priori voting power. The seriousness of this observation depends to some extent on the plausibility of the particular indices, but the fact that proportionality considerations completely overlook one crucial determinant of an actor's influence in voting bodies, viz. the decision rule, is a basic flaw that can be established without a commitment to any particular voting power index.

All voting bodies resort to a decision rule of some kind. The most common ones indicate the minimum size of the support a motion has to get in order to be passed. If the influence over the outcomes, i.e. the acts of the collective body, has anything to do with getting one's will expressed in the acts passed, then it would seem counterintuitive to say that a group with 10 % of the seats in a legislature has 10 % of the influence over the legislation if the rest of the body consists of one group with 90 % of the seats and if the decision rule is, say, 51 %, i.e. 51 % of the votes are needed to pass a motion and each seat is entitled to one vote. The smaller group simply cannot control any part of the legislation under these circumstances. Similarly if the decision rule is 51 % and the share of seats of a group increases from 49 % to 51 %, the increase in its influence would seem to be much larger than the difference between its seat shares would indicate. So it seems that the desire for proportionality without due attention to the decision rules leads to unsatisfactory outcomes.

The power indices are defined explicitly as functions of the decision rules. We shall focus here on two power indices, viz. the Shapley-Shubik index and the Banzhaf one (see Shapley and Shubik, 1954; Banzhaf, 1965). These are not doubt the most commonly used ones even though others have been discussed in the literature (see Deegan and Packel, 1979; Holler, 1982; Johnston, 1978; Laver, 1978). In the following introduction to power indices we shall first make use of Allingham's (1975) presentation.

Consider a simple n-person voting game, i.e. a game in which every subset of players is either winning or losing. In game-theoretic terminology a simple game is characterized by the following function defined for all subsets or coalitions S of the player set N: $v(S) = 1$ if S is winning and $v(S) = 0$, otherwise. The function v is called the characteristic function of the voting game. The game can then be defined as a pair (N,W) where W is the set of winning coalitions, i.e. $W \subset 2^N - \emptyset$ such that for any set T of players, $S \in W$ implies that $S \cup T \in W$.

Now a voter i is called dummy if he cannot benefit any coalition of voters by joining it, i.e.

i is dummy if $S \in W$ implies $S - \{i\} \in W$.

Two voters i and j are called symmetric if their benefit to any coalition is the same, that is, i and j are symmetric if for any S such that $i,j \notin S$: $S \cup \{i\} \in W$ iff $S \cup \{j\} \in W$.

The simple voting game (N,W) can be represented by a vote vector $x \in R^n$ and a quota number a if

$S \in W$ iff $\Sigma_{i \in S} x_i > a$,

where x_i is the number of votes of i'th voter.

Finally, if π is a permutation of voter set N, the permuted game (N, π W) is defined as the permutation of winners:

$\pi W = \{\pi S| S \in W\}$ where $\pi S = \{\pi i| i \in S\}$.

After these definitions the concept of power index can be introduced as follows (see Allingham, 1975, p. 296):

A power index for (N,W) is a probability vector g(N,S) satisfying four conditions:

1. $(\forall \pi)\, g_{\pi i}(N, \pi W) = g_i(N,W)$, i.e. the power index value of a player is not affected by the permutation of players. Only their numbers change.

2. The power index value of a dummy player is zero.

3. Symmetric players have identical power index values, and

4. If (N,W) can be given a vote vector - quota number representation, then for any two voters i and j: $x_i > x_j$ implies that i's power index value is at least as large as that of j.

Both the Shapley-Shubik and the Banzhaf index have been axiomatized. This means that the conditions both necessary and jointly sufficient for a voting power measure to be the index in question have been presented (see Harsanyi, 1977, pp. 215-226; Dubey and Shapley, 1979). Both of these indices satisfy conditions 1.-4. above. But since these indices are non-equivalent, conditions 1.-4. although necessary cannot be sufficient. What differentiates the two power indices is the fifth condition which for the Shapley-Shubik index is the following:

5. Additivity. Let the game G = (N,W) be a composite game consisting of two constituent games G_1 and G_2 in the following sense: $v(S) = v_1(S) + v_2(S)$, for all $S \subseteq N$. Here v is the characteristic function of the game G and v_1 (v_2, respectively) is the characteristic function of game G_1 (G_2). The additivity condition requires that the power index value of any player i in G is the sum of his power index values in G_1 and G_2.

For the Banzhaf index the differentiating axiom is the following.

5'. For any two games G_1 and G_2 with characteristic functions v_1 and v_2, respectively, there is a real valued function f such that for all $S \subseteq N$:

$$f(\max(v_1(S), v_2(S))) + f(\min(v_1(S), v_2(S))) = f(v_1(S)) + f(v_2(S)).$$

Moreover, $f(v(S)) = h(v(S))$ for all $S \subseteq N$, where

$h(v(S)) = \Sigma_{i \in S}\, h_i(v) =$ the number of swings for voter i. The swing for voter i is a pair (T,T-{i}) where T is a winning coalition and T-{i} is non-winning. Thus the number of swings for a voter is the number of times his entrance changes a coalition from a non-winning to a winning one.

The axiomatic characterizations of the power indices are not very illuminating when one tries to justify their use as measures

FROM COMMITTEES TO ELECTIONS

of a priori voting power. The common axioms 1.-4. are intuitively plausible, though, and bring out some of their intuitive background. The differentiating axioms 5. and 5'., on the other hand, are fairly technical and by no means obvious. The explicit definitions of the indices are, therefore, helpful.

$$F_i^{Sh} = \sum_{\substack{i \in S \\ S \subseteq N}} \frac{|S-\{i\}|! \, |N-S|!}{N!} [v(S) - v(S-\{i\})]$$

Here F_i^{Sh} is the so-called Shapley value of player i in a game with the characteristic function v. When we are dealing with simple games the value of v(S) can be either 0 or 1. In such games F_i^{Sh} is called i's Shapley-Shubik index value. It consists of i's contributions to various coalitions S weighted by a term which is constant for any coalition of the same size.

The (normalized) Banzhaf index value of voter i, in turn, is defined as

$$F_i^{Bz} = \frac{\sum_{\substack{i \in S \\ S \subseteq N}} [v(S) - v(S-\{i\})]}{\sum_{i \in N} \sum_{S \subseteq N} [v(S) - v(S-\{i\})]}$$

(Brams and Affuso, 1976).
The numerator of this index counts the number of times i's presence has been crucial or "critical" in various coalitions as obviously the expression in brackets is non-zero, i.e. one, only in cases where the removal of i from a winning S makes the remaining coalition S-{i} losing. The denominator, in turn, is the number of all critical presences or swings in the simple voting game.

Both indices thus pay attention to a voter's ability to change a coalitions's status from winning to losing. The difference is that the Shapley-Shubik index uses weights in summing up a voter's contributions to various coalitions, whereas the Banzhaf index puts an equal emphasis on each swing. The indices are, thus, non-equivalent although based on partly similar motivations. The issue of which index - if either - is the right one or the true measure of a priori voting power has been widely discussed in the literature. It seems that in different institutional environments different indices are appropriate (see Brams and Lake, 1978; Nurmi, 1980b; Nurmi, 1982; Riker and Shapley, 1968; Straffin, 1977). More important than the precise values of indices is the order of magnitude in which the indices can deviate from proportional seat or vote distributions. Let us consider first a fictitious example.

Example 13.2.

The support to parties A, B and C is 40 %, 35 % and 25 %, respectively. In a 100-member legislature perfect proportionality would thus call for a seat distribution of 40 seats to A, 35 to B, and 25 seats to C. Assume that the decision rule used in the legislature is 61. The swings for A are then:

(AB,B), (AC,C), (ABC,BC).

The only swings for B and C are (AB,A) and (AC,A), respectively. The Banzhaf index values are then:

$$F_A^{Bz} = 3/5, \ F_B^{Bz} = 1/5, \ F_C^{Bz} = 1/5.$$

The Shapley-Shubik index values, in turn, are

$$F_A^{Sh} = \frac{1!1!}{3!} + \frac{1!1!}{3!} + \frac{2!0!}{3!} = 2/3$$

$$F_B^{Sh} = F_C^{Sh} = \frac{1!}{3!} = 1/6.$$

Consequently, with 40 % of the seats or votes A would have 60 % of the voting power if the latter is measured by the Banzhaf index. Similarly, the parties B and C would have 20 % of the voting power each despite the fact that their shares of seats differ considerably from this figure and from each other. The same kind of phenomenon in an even more extreme form occurs when the Shapley-Shubik index is used as the measure of a priori voting power: A's 40 % of the seats corresponds now to about 66 % of the total a priori voting power, while B and C have only about 16 % of the total voting power each.

The differences between a priori voting power distributions and seat distributions can, thus, be considerable and certainly of an entirely different order of magnitude than the differences between, say, the Hamilton's and d'Hondt's apportionments. The above example is far from extreme in this respect. Indeed, if the decision rule in the above voting body were 70, C with its 25 seats would have no a priori voting power because there would be no swings for a party of that size.

Voting power considerations have obvious implications for institutional design: if it is the voting power and not merely the seats one wishes to distribute proportionally among the groups, then new methods of seat or vote allocation, given the support distribution, will have to be resorted to. However, there is another way of achieving identity between the a priori voting power and relative support, viz. that of manipulating the decision rules (see Nurmi, 1982). This method is largely of theoretical interest as its crux lies in using randomized decision rules. These, in turn, may be somewhat difficult to "sell" to the practitioners. Theoretically, however, their use would solve the problem as it is known that the relative seat distribution is identical with the expected value of the a priori voting power in the Shapley-Shubik sense when each decision rule from $n/2$ to n is used with equal probability (Laakso, 1975). Thus, if this index were accepted as the right measure of voting power, one could guarantee that the distribution of the latter would be identical with the seat distribution by using each of the above decision rules with the same probability.

Another implication for institutional design is that in some cases quite counterintuitive phenomena may occur as a consequence of seemingly natural institutional changes. Brams and Affuso (1985) give some examples from the Council of Ministers of the European Community. In 1958-1973 Luxembourg had one vote out of 17 in the Council where the decision rule at the time was 12. However, since each of the other member states had an even number of votes, there was no swing for Luxembourg. Hence it had no voting power. In 1973-1981 the total number of votes in the body was 58 (three new member states had been admitted and the previous members were given more votes than formerly). Luxembourg now had 2 votes and the decision rule was 41. Luxembourg's Banzhaf index value now became 0.016, i.e. larger than zero even though its relative share of votes 2/58 was smaller than before (1/17).

Another peculiar fact pertains to the Council from 1981 onwards. In this body a new member had been admitted with 5 votes. The total number of votes now became 63 and the new decision rule was adopted, viz. 45. In this body there are three member states with the same Banzhaf index value and yet not all with the same number of votes: Luxembourg with 2 votes has the same Banzhaf index value as Denmark and Ireland with 3 votes each. The votes obviously are not simply translatable into a priori voting power, an observation that should interest people deciding how to distribute votes fairly among the members of voting bodies.

Taken more generally the preceding observations seem to suggest that instead of trying merely to make the voting body as similar to the population in toto as possible one should bear in mind that the collective choices are made using some decision rules. These rules often make the a priori voting power distributions very different from the seat or vote distributions. By using several decision rules with given probabilities one can, however, render any seat distribution identical with the a priori voting power distribution when the latter is defined by means of expected values of the power indices (see Nurmi, 1982; Holler, 1985; Berg and Holler, 1984). This possibility should perhaps be taken into account when the decision rules of voting bodies are pondered upon.

CHAPTER 14

CONCLUSIONS

The preceding chapters show that even within the restricted set of voting procedures considered in the present study there is a wide range of variation with respect to performance on various criteria of goodness. None of the procedures satisfies all the criteria used above, but all have some properties to recommend them. There are at least two possibilities in terms of which to assess these findings:

1) to conclude that no procedure is good enough for all purposes and, hence, we should revise our ideas of popular choices so that their results are viewed as nearly random and certainly more or less accidental, or

2) to conclude that the differences in performances should be taken into account in choosing procedures for use in various settings.

I find the latter conclusion more plausible as the former would completely ignore the considerable differences between methods. Still, I would agree that our intuitions about what the procedures purportedly do should take account of their theoretical properties. Thus, when I advocate a sort of relativism in the use of procedures it is because the preceding discussion shows that the institutional setting in which they are used together with the procedures determines the extent to which various democratic ideals can be fulfilled.

The first conclusion reflecting this relativism concerns the kind of input one utilizes in making social choices. We have considered mostly the "traditional" type of input, i.e. the individual ordinal preferences. Surely, most of the procedures discussed above do not require such elaborate information, but it seems that the comparative performance of the methods is most easily evaluated when the individual preferences are known. Thus, regardless of which particular procedure is used in making social choices, one should ask the individuals to give their preference orderings. This is important because the possible anomalies are more easily discernible when the preference orders are known. Hence experimentation with various methods becomes possible.

Moreover, we have in Chapters 10 and 11 discussed the implications of asking individuals to reveal their preferences more or less fully than by giving their preference orderings. The former alternative, i.e. asking for more information, would indeed seem plausible provided that the individuals are genuinely disinterested in the issues to be decided upon. Under such circumstances they would seem less likely to misrepresent their individual utilities or/and other input data used in preference aggregation. In typical political settings the assumption that the voters are disinterested is, however, unlikely to hold.

If, on the other hand, people are asked to reveal few of their top-most alternatives with or without indicating their preferences, the performance of the preference aggregation methods - when assessed in terms of the entire preference orderings - becomes no better but often worse than when dealing with whole preference schedules. Furthermore, the results based on individual choice functions reveal that problems similar to the Arrow impossibility theorem can be encountered in such settings.

Secondly, the observation that each procedure considered in the preceding is vulnerable to strategic manipulation should not overshadow the important differences between the procedures as far as various types of strategic manoeuverings are concerned. For instance, the misrepresentation of individual preferences can only be beneficial to a voter if he knows enough about the preferences of others. To some extent this knowledge can be concealed by various institutional or normative arrangements.

The seriousness of another form of strategic behaviour, viz. agenda-manipulation, can also be restricted by arrangements dictating agenda-building. Moreover, some procedures are inherently difficult to manipulate even with an extensive knowledge of other people's preferences. When no such arrangements can be made, it would seem that the approval voting is the optimal method when primary attention is on the agenda-manipulation possibilities. If, on the other hand, individual preference misrepresentation is to be feared, then Hare's, Coombs' or plurality runoff procedure would seem appropriate. In general, our discussion in Chapter 9 shows that there is a considerable variation among methods as far as the vulnerability to various forms of strategic behaviour is concerned.

Thirdly, the problem of cyclic majorities haunts most directly the amendment procedure. Its solution in that setting would simply be for the voters to report their full preference orderings over the alternatives whereupon the possible cycle could immediately be spotted. Similarly, the possible Condorcet winner could equally well be identified. Otherwise, the procedure could be implemented in the usual way, i.e. by constructing an agenda with motions,

CONCLUSIONS

amendments, and status quo alternatives placed in their proper positions. The only difference from the present practice is that the determination of the winner would need some computation easily performed by computers.

Fourthly, the problem of making social choices out of a set of alternatives is basically the same regardless of whether one is choosing a collective body of individuals or a policy alternative. But representative democracy has an additional problem, viz. that of attaining proportionality when it is generally deemed desirable. Because decisions in collective bodies are made in accordance with fixed decision rules or qualified majorities one should carefully distinguish proportional voting power from proportional seat allocation when the support is given. As stated in the preceding chapter the problems encountered in making these two distributions identical can in principle be solved by resorting to randomized decision rules.

In sum, the investigation of the institutional and normative setting in which the procedures are supposed to operate is most helpful in deciding which particular procedure to adopt. In experimenting with the procedures the relative frequencies with which various undesirable features occur could be found out. While the theoretical work undertaken above can establish which problems may possibly be encountered in using various procedures, their practical application alone, using the entire preference profiles as input, can in the end determine how often these phenomena may occur.

BIBLIOGRAPHY

Aizerman, M. A.: 1984, *New Problems in the General Choice Theory*, manuscript, Institute of Control Sciences, Moscow.
Aizerman, M. A., and F. T. Aleskerov: 1983, 'Local Operators in Models of Social Choice', *Systems & Control Letters* **3**, 1-6.
Aizerman, M. A., and A. V. Malishevski: 1981, 'General Theory of Best Variants Choice: Some Aspects', *IEEE Transactions on Automatic Control* **AC-26**, 1030-1040.
Allingham, M. G.: 1975, 'Economic Power and Values of Games', *Zeitschrift für Nationalökonomie* **35**, 293-299.
Arrow, K. J.: 1959, 'Rational Choice Functions and Orderings', *Economica* **26**, 121-127.
Arrow, K. J.: 1963, *Social Choice and Individual Values*, Wiley, New York, 2nd ed.
Balinski, M. L., and H. P. Young: 1982, *Fair Representation*, Yale University Press, New Haven.
Banks, J. S.: 1985, 'Sophisticated Voting Outcomes and Agenda Control', *Social Choice and Welfare* **4**, 295-306.
Banzhaf, J. F.: 1965, 'Weighted Voting Doesn't Work: A Mathematical Analysis', *Rutgers Law Review* **19**, 329-330.
Barbera, S.: 1977, 'The Manipulation of Social Choice Mechanisms That Do Not Leave "Too Much" to Chance', *Econometrica* **45**, 1573-1588.
Berg, S., and M. J. Holler: 1984, *Randomized Decision Rules in Voting Games*, mimeo.
Black, D.: 1958, *Theory of Committees and Elections*, Cambridge University Press, Cambridge.
Bogdanor, V.: 1983, 'Introduction', in V. Bogdanor and D. Butler (eds.), *Democracy and Elections*, Cambridge University Press, Cambridge, pp. 1-19.
Brams, S. J.: 1976, *Paradoxes in Politics: An Introduction to the Non-obvious in Political Science*, Free Press, New York.
Brams, S. J.: 1982, 'The AMS Nomination Procedure Is Vulnerable to "Truncation of Preferences"', *Notices of the American Mathematical Society* **29**, 136-138.
Brams, S. J.: 1983, *Superior Beings: If They Exist, How Would We Know?*, Springer-Verlag, New York.
Brams, S. J., and P. J. Affuso: 1976, 'Power and Size: A New Paradox', *Theory and Decision* **7**, 29-56.

Brams, S. J., and P. J. Affuso: 1985, 'New Paradoxes of Voting Power in the EC Council of Ministers', *Electoral Studies* **4**, 135-139.

Brams, S. J., and P. C. Fishburn: 1978, 'Approval Voting', *American Political Science Review* **72**, 831-847.

Brams, S. J., and P. C. Fishburn: 1983 (a), *Approval Voting*, Birkhäuser, Boston.

Brams, S. J., and P. C. Fishburn: 1983 (b), 'America's Unfair Elections', *The Sciences* **6**, 28-34.

Brams, S. J., and M. Lake: 1978, 'Power and Satisfaction in a Representative Democracy', in P. Ordershook (ed.), *Game Theory and Political Science*, New York University Press, New York, pp. 529-562.

Brams, S. J., and D. Wittman: 1981, 'Nonmyopic Equilibria in 2 x 2 Games', *Conflict Management and Peace Science* **6**, 39-62.

Campbell, D. E.: 1982, 'On the Derivation of Majority Rule', *Theory and Decision* **14**, 133-140.

Carter, C.: 1982, 'Some Properties of Divisor Methods for Legislative Apportionment and Proportional Representation', *American Political Science Review* **76**, 575-584.

Chamberlin, J., and M. Cohen: 1978, 'Towards Applicable Social Choice Theory: A Comparison of Social Choice Functions under Spatial Model Assumptions', *American Political Science Review* **72**, 1241-1256.

Chamberlin, J., J. L. Cohen, and C. H. Coombs: 1984, 'Social Choice Observed: Five Presidential Elections of the American Psychological Association', *The Journal of Politics* **46**, 479-502.

Cohen, L.: 1979, 'Cyclic Sets in Multidimensional Voting Models', *Journal of Economic Theory* **20**, 1-12.

Colman, A., and I. Pountney: 1978, 'Borda's Voting Paradox: Theoretical Likelihood and Electoral Occurrences', *Behavioral Science* **23**, 15-20.

Coughlin, P.: 1982, 'Pareto Optimality of Policy Proposals with Probabilistic Voting', *Public Choice* **39**, 427-433.

Deegan, J., Jr., and E. W. Packel: 1979, 'A New Index of Power for Simple n-Person Games', *International Journal of Game Theory* **7**, 113-123.

DeGrazia, A.: 1953, 'Mathematical Derivation of an Election System', *Isis* **44**, 42-51.

Doron, G.: 1979, 'The Hare System is Inconsistent', *Political Studies* **27**, 283-286.

Dubey, P., and L. S. Shapley: 1979, 'Mathematical Properties of the Banzhaf Power Index', *Mathematics of Operations Research* **4**, 99-130.

Dutta, B., and P. K. Pattanaik: 1978, 'On Nicely Consistent Voting Systems', *Econometrica* 46, 163-170.
Dyer, J. S., and R. E. Miles, Jr.: 1976, 'An Actual Application of Collective Choice Theory to the Selection of Trajectories for Mariner Jupiter/Saturn 1977 Project', *Operations Research* 24, 220-244.
Elster, J.: 1978, 'Kjernekraft - Noen politiske og beslutsteoretiske momenter', in *Riskvärdering*, Industridepartementet, Stockholm, pp. 1-61.
Enelow, J. M., and M. J. Hinich: 1984, 'Probabilistic Voting and the Importance of Centrist Ideologies in Democratic Elections', *The Journal of Politics* 46, 459-478.
Farquharson, R.: 1969, *Theory of Voting*, Yale University Press, New Haven.
Feldman, A.: 1980, *Welfare Economics and Social Choice Theory*, Martinus Nijhoff, Boston.
Ferejohn, J. A., and D. M. Grether: 1974, 'On a Class of Rational Social Decision Procedures', *Journal of Economic Theory* 8, 471-482.
Fishburn, P. C.: 1973, *The Theory of Social Choice*, Princeton University Press, Princeton.
Fishburn, P. C.: 1974, 'Paradoxes of Voting', *American Political Science Review* 68, 537-546.
Fishburn, P. C.: 1977, 'Condorcet Social Choice Functions', *SIAM Journal on Applied Mathematics* 33, 469-489.
Fishburn, P. C.: 1979, 'Symmetric and Consistent Aggregation with Dichotomous Voting', in J.-J. Laffont (ed.), *Aggregation and Revelation of Preferences*, North-Holland, Amsterdam, pp. 201-218.
Fishburn, P. C.: 1982, 'Monotonicity Paradoxes in the Theory of Voting', *Discrete Applied Mathematics* 4, 119-134.
Fishburn, P. C.: 1983, 'Dimensions of Election Procedures: Analyses and Comparisons', *Theory and Decision* 15, 371-397.
Fishburn, P. C., and S. J. Brams: 1981, 'Approval Voting, Condorcet's Principle, and Runoff Elections', *Public Choice* 36, 89-114.
Fishburn, P. C., and S. J. Brams: 1983, 'Paradoxes of Preferential Voting', *Mathematics Magazine* 56, 207-214.
Fishburn, P. C., and S. J. Brams: 1984, 'Manipulability of Voting by Sincere Truncation of Preferences', *Public Choice* 44, 397-410.
Gärdenfors, P.: 1976, 'Manipulation of Social Choice Functions', *Journal of Economic Theory* 13, 217-228.
Gärdenfors, P.: 1977, 'A Concise Proof of a Theorem on Manipulation of Social Choice Functions', *Public Choice* 32, 137-142.

BIBLIOGRAPHY

Gibbard, A.: 1973, 'Manipulation of Voting Schemes', *Econometrica* 41, 587-601.
Green, J. R., and J.-J. Laffont: 1979, *Incentives in Public Decision-Making*, North-Holland, Amsterdam.
Greenberg, J.: 1979, 'Consistent Majority Rule over Compact Sets of Alternatives', *Econometrica* 47, 627-636.
Hansson, B., and H. Sahlquist: 1976, 'A Proof Technique for Social Choice with Variable Electorate', *Journal of Economic Theory* 13, 193-200.
Hardin, R.: 1971, 'Collective Action as an Agreeable n-Prisoner's Dilemma', *Behavioral Science* 16, 472-481.
Hardin, R.: 1982, *Collective Action*, Johns Hopkins Press, Baltimore.
Hare, T.: 1861, *The Election of Representatives, Parliamentary and Municipal: A Treatise*, Longman, Green, London.
Harsanyi, J. C.: 1977, *Rational Behavior and Bargaining Equilibrium in Games and Social Situations*, Cambridge University Press, Cambridge.
Hinich, M. J., J. O. Ledyard, and P. C. Ordeshook: 1973, 'A Theory of Electoral Equilibrium', *The Journal of Politics* 35, 154-193.
Holler, M. J.: 1982, 'Forming Coalitions and Measuring Voting Power', *Political Studies* 30, 262-271.
Holler, M. J.: 1985, 'Strict Proportional Power in Voting Bodies', *Theory and Decision* 19, 249-258.
Johnston, R. J.: 1978, 'On the Measurement of Power: Some Reactions to Laver', *Environment and Planning* 10A, 907-914.
Kelly, J. S.: 1978, *Arrow Impossibility Theorems*, Academic Press, New York.
Kelsey, D.: 1984, *Topics in Social Choice*, D. Phil. Thesis, Oxford.
Kelsey, D.: 1985, 'The Liberal Paradox: A Generalization', *Social Choice and Welfare* 1, 245-250.
Kim, K. H., and F. W. Roush: 1980, *Introduction to Mathematical Consensus Theory*, Marcel Dekker, New York.
Kramer, G. H.: 1973, 'On a Class of Equilibrium Conditions for Majority Rule', *Econometrica* 41, 285-297.
Kramer, G. H.: 1977, 'A Dynamical Model of Political Equilibrium', *Journal of Economic Theory* 16, 310-334.
Laakso, M.: 1975, *The Finnish Parliament as a Coalition and Power Relation Structure*, Acta Politica, Fasc. IX, Helsinki (in Finnish).
Larsson, B.: 1983, *Basic Properties of Majority Rule*, University of Lund, Department of Economics.
Laver, M.: 1978, 'The Problems of Measuring Power in Europe', *Environment and Planning* 10 A, 901-906.
Laver, M.: 1981, *The Politics of Private Desires*, Penguin Books, Harmondsworth.

Lehrer, K., and C. Wagner: 1981, *Rational Consensus is Science and Society*, D. Reidel, Dordrecht.
McKelvey, R. D.: 1976, 'Intransitivities in Multidimensional Voting Models and Some Implications for Agenda Control', *Journal of Economic Theory* **12**, 472-482.
McKelvey, R. D.: 1979, 'General Conditions for Global Intransitivities in Formal Voting Models', *Econometrica* **47**, 1085-1112.
McKelvey, R. D., and R. G. Niemi: 1978, 'A Multistage Game Representation of Sophisticated Voting for Binary Procedures', *Journal of Economic Theory* **18**, 1-22.
May, K. O.: 1952, 'A Set of Independent, Necessary and Sufficient Conditions for Simple Majority Decision', *Econometrica* **20**, 680-684.
Merrill, S.: 1984, 'A Comparison of Efficiency of Multicandidate Electoral Systems', *American Journal of Political Science* **28**, 23-48.
Miller, N. R.: 1980, 'A New Solution Set for Tournaments and Majority Voting: Further Graph-Theoretical Approaches to the Theory of Voting', *American Journal of Political Science* **24**, 68-96.
Miller, N. R.: 1983, 'Pluralism and Social Choice', *American Political Science Review* **77**, 734-747.
Nakamura, K.: 1978, 'The Vetoers in a Simple Game with Ordinal Preferences', *International Journal of Game Theory* **8**, 55-61.
von Neumann, J., and O. Morgenstern: 1944, *Theory of Games and Economic Behavior*, Princeton University Press, Princeton.
Niemi, R. G.: 1983, 'Why So Much Stability?', *Public Choice* **41**, 261-270.
Nurmi, H.: 1980 (a), 'Majority Rule: Second Thoughts and Refutations', *Quality and Quantity* **14**, 743-765.
Nurmi, H.: 1980 (b), 'Game Theory and Power Indices', *Zeitschrift für Nationalökonomie* **40**, 35-58.
Nurmi, H.: 1982, 'The Problem of the Right Distribution of Voting Power', in M. J. Holler (ed.), *Power, Voting, and Voting Power*, Physica-Verlag, Würzburg, pp. 203-212.
Nurmi, H.: 1983 (a), 'Voting Procedures: A Summary Analysis', *British Journal of Political Science* **13**, 181-208.
Nurmi, H.: 1983 (b), 'Past Masters and Their Modern Followers', in I. Heiskanen, and S. Hänninen (eds.), *Exploring the Basis of Politics*, Finnpublishers, Tampere, pp. 109-131.
Nurmi, H.: 1983 (c), 'On Riker's Theory of Political Succession', *Scandinavian Political Studies* **6**, 177-194.
Nurmi, H.: 1985, 'On Some Properties of the Lehrer-Wagner Model for Reaching Rational Consensus', *Synthese* **62**, 13-24.

Nurmi, H.: 1984 (a), 'On Taking Preferences Seriously', in D. Anckar and E. Berndtson (eds.), *Essays in Democratic Theory*, Finnpublishers, Tampere, pp. 81-104.
Nurmi, H.: 1984 (b), 'On the Strategic Properties of Some Modern Methods of Group Decision Making', *Behavioral Science* **29**, 248-257.
Nurmi, H., and E. Lagerspetz: 1984, 'Observations on the Finnish Electoral System', in D. Anckar and E. Berndtson (eds.), *Essays in Democratic Theory*, Finnpublishers, Tampere, pp. 105-123.
Olson, M.: 1965, *The Logic of Collective Action*, Harvard University Press, Cambridge, Mass.
Pattanaik, P. K.: 1971, *Voting and Collective Choice*, Cambridge University Press, Cambridge.
Pattanaik, P. K.: 1978, *Strategy and Group Choice*, North-Holland, Amsterdam.
Plott, C.: 1973, 'Path Independence, Rationality, and Social Choice', *Econometrica* **41**, 1075-1091.
Plott, C.: 1976, 'Axiomatic Social Choice Theory: An Overview and Interpretation', *American Journal of Political Science* **20**, 511-596.
Rawls, J.: 1971, *A Theory of Justice*, Harvard University Press, Cambridge, Mass.
Richelson, J. T.: 1978, 'A Comparative Analysis of Social Choice Functions, II', *Behavioral Science* **23**, 38-44.
Richelson, J. T.: no date, *Majority Rule and Collective Choice*, mimeo.
Riker, W. H.: 1982, *Liberalism against Populism: A Confrontation between the Theory of Democracy and the Theory of Social Choice*, W. H. Freeman, San Francisco.
Riker, W. H., and P. C. Ordeshook: 1973, *An Introduction to Positive Political Theory*, Prentice-Hall, Englewood Cliffs.
Riker, W. H., and L. S. Shapley: 1968, 'Weighted Voting: A Mathematical Analysis for Instrumental Judgements', in R. Pennock and J. W. Chapman (eds.), *Nomos X: Representation*, Atherton Press, New York, pp. 199-216.
Roberts, K. W.: 1980, 'Interpersonal Comparability and Social Choice Theory', *Review of Economic Studies* **XLVII**, 421-439.
Satterthwaite, M.: 1975, 'Strategy-Proofness and Arrow's Conditions', *Journal of Economic Theory* **10**, 187-217.
Schofield, N.: 1983 (a), 'Classification of Voting Games on Manifolds', California Institute of Technology, *Social Science Working Paper* **488**.

Schofield, N.: 1983 (b), 'Equilibria in Simple Dynamic Games', in Pattanaik, P. K., and M. Salles (eds.), *Social Choice and Welfare*, North-Holland, Amsterdam, pp. 269-284.
Schofield, N.: 1984, 'Social Equilibrium and Cycles on Compact Sets', *Journal of Economic Theory* **33**, 59-71.
Schwartz, T.: 1972, 'Rationality and the Myth of the Maximum', *Nous* **6**, 97-117.
Sen, A. K.: 1966, 'A Possibility Theorem on Majority Decisions', *Econometrica* **34**, 491-499.
Sen, A. K.: 1970, *Collective Choice and Social Welfare*, Holden-Day, San Francisco.
Sen, A. K.: 1976, 'Liberty, Unanimity and Rights', *Economica* **43**, 217-245.
Sen, A. K.: 1983, 'Liberty and Social Choice', *The Journal of Philosophy* **LXXX**, 5-28.
Sen, A. K.: 1986, 'Social Choice Theory', in K. J. Arrow, and M. D. Intriligator (eds.), *Handbook in Mathematical Economics*, Vol. III, North-Holland, Amsterdam, pp. 1073-1181.
Shapley, L. S., and M. Shubik: 1954, 'A Method of Evaluating the Distribution of Power in a Committee System', *American Political Science Review* **48**, 787-792.
Shepsle, K. A.: 1979, 'Institutional Arrangements and Equilibrium in Multidimensional Voting Models', *American Journal of Political Science* **23**, 27-59.
Shepsle, K. A., and B. R. Weingast: 1981, 'Structure-Induced Equilibrium and Legislative Choice', *Public Choice* **37**, 503-519.
Simpson, P. B.: 1969, 'On Defining Areas of Voter Choice', *Quarterly Journal of Economics* **LXXXIII**, 478-490.
Slutsky, S.: 1979, 'Equilibrium under α-Majority Voting', *Econometrica* **47**, 1113-1127.
Smith, J. H.: 1973, 'Aggregation of Preferences with Variable Electorate', *Econometrica* **41**, 1027-1041.
Straffin, P. D., Jr.: 1980, *Topics in the Theory of Voting*, Birkhäuser, Boston.
Suzumura, K.: 1983, *Rational Choice, Collective Decisions, and Social Welfare*, Cambridge University Press, Cambridge.
Taylor, M.: 1976, *Anarchy & Cooperation*, Wiley, New York.
Tideman, T. N., and G. Tullock: 'A New and Superior Process for Making Social Choices', *Journal of Political Economy* **84**, 1145-1159.
Ullman-Margalit, E.: 1977, *The Emergence of Norms*, Oxford University Press, Oxford.
Vickrey, W.: 1960, 'Utility, Strategy and Social Decision Rules', *Quarterly Journal of Economics* **LXXIV**, 507-535.

Weber, R. J.: 1977, 'Comparison of Voting Systems', *Cowles Foundation Discussion Paper*, No. **498** A.
Young, H. P.: 1974, 'An Axiomatization of Borda's Rule', *Journal of Economic Theory* **9**, 43-52.
Young, H. P.: 1975, 'Social Choice Scoring Functions', *SIAM Journal on Applied Mathematics* **28**, 824-838.

SUBJECT INDEX

acyclicity 9, 19, 21, 22, 23, 24
agenda-control 20, 110, 111, 126, 127
   -manipulability 20, 110, 111, 125, 126, 127
   -setter 20, 21, 163, 172, 174
Alabama paradox 182, 183
alternative vote (see Hare's procedure) 54
amendment procedure 14, 15, 16, 24, 31, 40, 41, 69, 71, 72, 77, 79, 86, 87, 88, 101, 104, 114, 126, 127, 131, 149, 161, 162, 170, 172, 192
anonymity 101, 112, 113, 156, 157
apportionment 181, 182, 183, 188, 195
approval voting 3, 56, 57, 58, 59, 60, 63, 73, 74, 78, 85, 86, 88, 100, 101, 103, 104, 105, 106, 116, 121, 123, 124, 125, 126, 127, 175, 176, 192, 195, 196
Arrow impossibility theorem 87, 98, 192
asymmetric relation 25
backward induction 167, 170
Banzhaf index (see power indices)
Bentham method 140
binary criteria 61
   methods 64, 74, 123
Black's procedure (method) 47, 62, 69, 73, 78, 79, 84, 132
Borda count 3, 31, 32, 33, 34, 37, 45, 46, 47, 49, 59, 60, 62, 63, 73, 74, 78, 79, 83, 84, 99, 100, 115, 116, 123, 133, 141, 152, 154, 173, 180
   paradox 48, 195
   score 32, 45, 46, 71, 73, 78, 84, 101, 115, 132, 133, 141
   winner 32, 37, 39, 46, 47, 63, 73, 78, 84, 132, 133
cardinality (of utility functions) 22, 29, 70, 139, 140, 144, 148
characteristic function (of a voting game) 185, 186, 187
citizens' sovereignty 88
closure 25, 28, 31
collective (or social) preference relation 8, 9, 10, 12, 13, 139, 149, 172
   rationality (see group rationality)
complete relations 6
concordance 157, 158, 159
Condorcet condition 39, 40, 41, 46, 49, 53, 54, 55, 56, 57, 60, 61, 79
   moderate 130
   strong 39, 40, 41, 42, 46, 47, 49, 52, 53, 56, 60, 130, 131
Condorcet loser 40, 41, 42, 45, 46, 47, 48, 49, 52, 53, 54, 55, 57, 60, 144, 145, 151, 175
   losing (or loser) criterion 41, 47, 49, 52, 54, 56
   paradox 12, 13, 15, 16, 19, 23, 25, 26, 28, 30, 31, 32, 34, 35, 104, 112, 114, 162

SUBJECT INDEX

winner 15, 16, 19, 21, 22, 26, 27, 28, 29, 33, 36, 38, 39, 40, 41, 42, 45, 46, 47, 48, 49, 50, 51, 52, 53, 54, 55, 57, 58, 59, 60, 62, 64, 69, 70, 72, 73, 75, 78, 83, 84, 88, 104, 112, 113, 122, 123, 126, 132, 150, 151, 155, 160, 162, 169, 170, 171, 172, 176, 177, 192

winning criterion 27, 39, 40, 41, 47, 48, 49, 52, 54, 55, 56, 58, 59, 61, 62, 64, 66, 69, 71, 73, 74, 101, 113, 114, 131, 132, 133, 150, 176

consensual weight 142, 143

consistency 92, 93, 94, 100, 101, 102, 103, 104, 106, 157, 177

ballot 104, 125

weak 94

Coombs' procedure (method) 55, 62, 64, 73, 77, 85, 103, 105, 117, 123, 133, 148, 173, 176

Copeland choice function 42

score 42, 43, 44, 72, 122, 131

winner 43, 44, 72, 122

Copeland's procedure (method) 42, 44, 45, 69, 72, 77, 83, 131

core 19, 20, 21, 23, 24, 26, 39, 40, 41, 43, 44, 46, 47, 52, 53, 130, 166

cyclic majority 12, 14, 16, 17, 24, 26, 27, 35, 36, 37, 64, 192

decision rule 22, 82, 184, 185, 188, 189, 190, 193, 194, 201

deterministic voting 165, 166

divisor method 182, 183, 195

Dodgson winner 27, 28, 70, 122

Dodgson's procedure (method) 26, 27, 28, 29, 31, 44, 35, 36, 52, 53, 55, 56, 69, 70, 71, 73, 74, 74, 83, 84, 104, 121, 122, 123, 131, 132

dummy player 185, 186

eliminative (or elimination) procedure 72, 76, 88

Euclidean distance 18, 19

expected utility property 136, 137, 138

free rider 108, 109

Gärdenfors' theorem 113, 114

Gibbard-Satterthwaite theorem 110, 112, 117

group (or collective) rationality 89, 91, 92

Hamilton's method (formula) 182, 183

Hare winner 54, 105, 117

Hare's procedure (method) 54, 77, 85, 102, 117, 128, 173

heritage 97, 98, 100, 104, 105, 106, 107, 127, 145, 157, 158, 159, 176, 177

d'Hondt's method 182

house-monotonicity 181, 182

ideal point 17, 18, 19, 20, 21

indecisiveness 24

independence

of irrelevant alternatives (IIA) 98, 99, 145

of rejecting outcast variants (IRO) 98, 99, 100, 158, 159

indifference curve 19, 20, 22

relation 6

individual choice function 155, 156, 157, 159, 192

interpersonal comparability (of utilities) 136, 138, 139, 140, 143, 199

inter-profile approach 93

properties 93, 99

intra-profile approach 93

criteria 93

properties 94, 97, 98, 100, 104

invariance transformations 139
inverse Condorcet criterion 40 (see Condorcet losing criterion)
k-majority operators 159
Lehrer-Wagner method 141, 142, 146
lexicographic preference 143
leximin rule 143, 144, 145
liberal paradox 87, 89, 90, 91, 197
locality (of choice functions) 156, 159
majority preference cycle 13
    winning criterion 62, 63, 64, 145
manipulability 109, 110, 111, 112, 117, 118, 125, 144, 172, 174, 196
manipulation hierarchy 119
maximality 25, 28, 31
maximin method 29, 30, 31, 33, 53, 56, 69, 78, 84, 104, 122, 131
methodological holists 5
    individualism 5
    individualists 5
minimal liberty 87, 89, 91
minmax number 21, 22, 29
monotonicity 66, 67, 68, 69, 70, 71, 72, 73, 74, 75, 76, 77, 78, 79, 80, 81, 87, 88, 93, 145, 156, 175, 176, 181, 182, 183
    in prizes 115
    paradoxes 182, 196
    strict 139
    strong 67, 68, 69, 77, 78, 79, 82
Nakamura number 21, 22, 23, 24
Nanson's procedure (method) 46, 64, 70, 71, 73, 74, 84, 101, 104, 114
Nash equilibrium 161, 162
Nash method 141, 144, 146

negative voting 148
neutrality 15, 16, 101, 112, 113, 156, 157, 159
non-imposition 87, 88, 91
nonmyopic equilibrium 166, 167, 168, 195
non-ranked systems 59, 60
nontransitivity 12
quota violation 182, 183
    lower 182
    upper 182
paradox of new states 182
Pareto condition: 79, 81, 82, 84, 85, 86, 87
    conservative strong 82, 86
    conservative weak 82, 86
    strong 81, 82, 83, 84, 85, 86, 87
    weak 81, 82, 83, 85, 86, 87
Pareto optimality 79, 87, 88, 89, 90, 91, 92, 118, 175, 195
    set 163, 164, 166, 170
path-independence (PI) 97, 98, 100, 104, 105, 106, 126, 145
plurality runoff 2, 49, 50, 51, 52, 55, 56, 59, 61, 62, 74, 75, 77, 79, 84, 88, 101, 102, 105, 116, 120, 123, 124, 127, 132, 133, 162, 172, 180, 192
plurality voting (method) 2, 48, 49, 50, 57, 58, 59, 60, 74, 83, 105, 118
    winner 48
point counting method 27
population-monotonicity 181
positional criteria 73
    voting procedures 49, 64, 73, 74, 83, 159
power indices 185, 186, 190, 198
    Banzhaf index 186, 187, 188, 189, 196

SUBJECT INDEX

Shapley-Shubik index 185, 186, 187, 188
predictive space 165
preference profile 9, 10, 16, 17, 24, 33, 38, 42, 50, 67, 71, 73, 74, 75, 76, 79, 88, 90, 92, 93, 94, 97, 98, 99, 102, 103, 109, 110, 111, 112, 113, 118, 119, 120, 121, 122, 123, 124, 125, 130, 146, 161, 162, 170, 175, 176, 177, 193
preference relation 7, 8, 9, 10, 12, 13, 17, 22, 38, 110, 114, 135, 137, 139, 149, 155, 161, 172
    strict 6, 22, 25
    weak 6
preference revelation 109, 110, 114, 115, 116, 117, 118, 119, 120, 151, 163, 173
preferential voting 54, 196 (see Hare's procedure)
Prisoner's Dilemma 167, 168, 173, 174, 197
probabilistic voting 164, 165, 195, 196
public (or collective) goods provision 108, 109, 112
ranked systems 59, 60
rational consensus 142, 198, 199
responsiveness 66, 67
    non-negative 67
    positive 67, 77
scoring function 31, 33, 34, 101, 152, 201
Schwartz' procedure (method) 25, 26, 28, 29, 30, 31, 33, 35, 41, 64, 69, 72, 77, 86, 88, 104, 121, 122, 131, 149
sequential game 167
Shapley-Shubik index (see power indices)
Shapley value 187
simple game 185, 187, 198

simple voting game 185, 187
sincere truncation of preferences (or s. preference truncation) 125, 128, 129, 130, 131, 132, 133, 134, 196,
    voting 16, 38, 57, 58, 126, 151, 170, 171
single-peakedness 16, 17, 19, 23
single transferable vote 54, 180 (see Hare's procedure)
social choice function: 8, 9, 10, 60, 68, 88, 93, 101, 109, 110, 111, 113, 140, 157, 158, 180, 195, 196, 197, 199
    dictatorial 110
    nontrivial 110
    resolute 10
    single-valued 110
    universal 110
social preference cycle 90
    relation (see collective preference relation)
social welfare function 8, 9, 68, 88
sophisticated equivalence (SE) 169, 170, 171
    outcome 169
    voting 80, 168, 169, 170, 171, 194, 198
sovereignty (of choice functions) 156
straight-forward procedure 163
strategic misrepresentation of preferences 110, 119, 120
strategy-proof procedure 110
structure-induced equilibrium 163, 200
successive procedure 168
symmetric voters 185, 186
terminal outcome 167
top cycle set 26, 31, 35
transitivity of relations 6, 7, 172
unanimity 81, 83, 158, 200

strong 82
unrestricted domain 89, 91
von Neumann-Morgenstern utility 136, 138, 139, 140, 144, 146
voting equilibrium 23, 160, 170
weak axiom of revealed preference (WARP) 94, 95, 96, 97, 98, 100, 104, 105, 106, 107, 127, 145, 175, 176
weight matrix 142
winning coalition 22, 23, 185, 186

NAME INDEX

Affuso, P. 187, 189, 194, 195
Aizerman, M. 97, 98, 100, 155, 157, 158, 159, 194
Aleskerov, F. 155, 157, 158, 159, 194
Allingham, M. 185, 186, 194
Anckar, D. 199
Arrow, K. 8, 9, 24, 87, 94, 98, 110, 158, 159, 192, 194, 197, 198, 200
Balinski, M. 181, 182, 194
Banks, J. 113, 194
Banzhaf, J. 185, 186, 187, 188, 189, 194, 196
Barbera, S. 110, 194
Berg, S. 190, 194
Berndtson, E. 199
Black, D. 16, 23, 26, 32, 33, 45, 47, 59, 61, 62, 69, 73, 78, 79, 84, 101, 114, 124, 131, 132, 163, 194
Bogdanor, V. 178, 194
Borda, J-C. 3, 27, 31, 32, 33, 34, 36, 39, 45, 46, 47, 48, 49, 55, 59, 60, 61, 62, 63, 71, 73, 74, 78, 79, 83, 84, 99, 100, 101, 104, 115, 116, 123, 124, 132, 133, 141, 152, 154, 173, 180, 195, 201
Brams, S. ix, 56, 57, 102, 128, 130, 133, 166, 182, 187, 189, 194, 195, 196
Butler, D. 194
Campbell, D. 68, 195
Carter, C. 181, 195
Chamberlin, J. 58, 173, 195
Chapman, J. 199
Coakley, J. ix
Cohen, J. 195

Cohen, L. 20, 195
Cohen, M. 58, 195
Colman, A. 48, 195
Coombs, C. 55, 59, 60, 61, 62, 64, 73, 77, 85, 101, 103, 105, 116, 117, 123, 124, 133, 148, 173, 176, 192, 195
Copeland, A. 41
Coughlin, P. 164, 195
Deegan, J. 185, 195
DeGrazia, A. 32, 48, 195
Dodgson, C. 26, 27, 28, 29, 31, 33, 35, 36, 52, 53, 55, 56, 61, 69, 70, 71, 73, 74, 83, 84, 104, 121, 122, 123, 124, 131, 132
Doherty, G. ix
Doron, G. 54, 102, 195
Dubey, P. 186, 196
Dutta, B. 161, 196
Dyer, J. 141, 196
Elster, J. 89, 91, 196
Enelow, J. 165, 196
Farquharson, R. 27, 126, 168, 196
Feldman, A. 109, 110, 196
Ferejohn, J. 24, 196
Fishburn, P. 9, 27, 28, 29, 33, 39, 56, 57, 67, 70, 71, 72, 75, 76, 81, 82, 84, 102, 122, 130, 154, 163, 195, 196
Gärdenfors, P. 10, 110, 112, 113, 114, 197
Gibbard, A. 110, 112, 117, 197
Green, J. 109, 197
Greenberg, J. 21, 23, 197
Grether, D. 24, 196
Hansson, B. 33, 197
Hänninen, S. 199

NAME INDEX

Hardin, R. 108, 109, 197
Hare, T. 54, 55, 59, 60, 61, 62, 64, 73, 77, 85, 88, 101, 102, 105, 116, 117, 123, 124, 128, 133, 173, 176, 192, 195, 197
Harsanyi, J. 115, 136, 186, 197
Heiskanen, I. 199
Hilpinen, A. x
Hilpinen, R. ix
Hinich, M. 164, 165, 196, 197
Hintikka, J. ix
Holler, M. 185, 190, 194, 197, 198
Intriligator, M. 200
Johnston, R. 185, 197
Kelly, J. 68, 81, 110, 197
Kelsey, D. 7, 91, 197
Kim, K. 143, 197
Kramer, G. 19, 21, 24, 29, 84, 197
Laakso, M. 189, 197
Laffont, J-J. 109, 196, 197
Lagerspetz, E. 183, 184, 199
Lake, M. 187, 195
Larsson, B. 16, 197
Laver, M. ix, 109, 185, 197, 198
Ledyard, J. 197
Lehrer, K. 141, 142, 146, 198, 199
McKelvey, R. 20, 113, 160, 166, 168, 169, 171, 198
May, K. 67, 68, 77, 198
Malishevski, A. 97, 98, 100, 194
Merrill, S. 54, 58, 59, 60, 162, 198
Miles, R. 141, 196
Miller, N. 113, 172, 198
Morgenstern, O. 136, 138, 139, 140, 144, 146, 198
Nakamura, K. 21, 22, 23, 24, 198
Nanson, E. 46
von Neumann, J. 136, 138, 139, 140, 144, 146, 198

Niemi, R. ix, 113, 163, 168, 169, 171, 198
Nurmi, H. 28, 32, 33, 40, 57, 69, 86, 102, 105, 106, 131, 132, 133, 143, 161, 163, 183, 184, 187, 189, 190, 198, 199
Nurmi, K. x
Olson, M. 108, 109, 199
Ordeshook, P. 108, 138, 197, 199
Packel, E. 185, 195
Pattanaik, P. 67, 68, 110, 161, 196, 199, 200
Pennock, R. 199
Plott, C. 9, 68, 81, 95, 97, 199
Pountney, I. 48, 195
Rawls, J. 139, 199
Ray, D. ix, 45
Richelson, J. 27, 40, 42, 45, 104, 199
Riker, W. ix, 4, 12, 14, 15, 39, 87, 88, 108, 114, 138, 140, 160, 161, 170, 172, 187, 199
Roberts, K. 139, 199
Roush, F. 143, 197
Sahlquist, H. 33, 197
Salles, M. 200
Satterthwaite, M. 110, 112, 117, 200
Schofield, N. 21, 22, 23, 160, 166, 200
Schwartz, T. 25, 26, 28, 29, 30, 31, 33, 35, 41, 61, 64, 69, 72, 77, 79, 86, 88, 104, 121, 122, 124, 131, 149, 200
Sen, A. 9, 16, 25, 87, 89, 91, 97, 200
Shapley, L. 185, 186, 187, 188, 189, 196, 199, 200
Shepsle, K. 163, 172, 200
Shubik, M. 185, 186, 187, 188, 189, 200
Simpson, P. 29, 200
Slutsky, S. 21, 200

NAME INDEX

Smith, J. 70, 200
Straffin, P. 40, 54, 55, 71, 74, 187, 200
Suzumura, K. 91, 200
Taylor, M. 109, 200
Tideman, T. 109, 200
Tullock, G. 109, 200
Ullman-Margalit, E. 173, 201
Vickrey, W. 16, 201
Wagner, C. 141, 142, 146, 198, 199
Wakker, P. ix
Weber, R. 60, 201
Weingast, B. 163, 200
Wittman, D. 166, 195
Young, H. 34, 93, 100, 101, 181, 182, 194, 201